Interactive Speech Technology

Interactive Speech Technology: Human factors issues in the application of speech input/output to computers

Edited by

Christopher Baber

and

Janet M. Noyes

Taylor & Francis
Publishers since 1798

Published by Taylor & Francis
2 Park Square, Milton Park, Abingdon, Oxon, OX14 4RN
270 Madison Ave, New York NY 10016

Transferred to Digital Printing 2010

British Library Cataloguing in Publication Data

A catalogue record for this book is available from the British Library

ISBN 0-7484-0127-X

Cover design by Hybert Design & Type

Typeset by Euroset, 2 Dover Close, Alresford, Hampshire SO24 9PG

Publisher's Note
The publisher has gone to great lengths to ensure the quality of this reprint but points out that some imperfections in the original may be apparent.

Contents

Contributors

W. A. Ainsworth
Department of Communication and Neuroscience, Keele University
Keele, Staffs. ST5 5BG

J. L. Arnott
The Microcomputer Centre
Department of Mathematics and Computer Science, The University
Dundee DD1 4HN

C. Baber
Industrial Ergonomics Group,
School of Manufacturing and Mechanical Engineering, University of Birmingham
Birmingham B15 2TT

A. Y. Cairns
The Microcomputer Centre
Department of Mathematics and Computer Science, The University
Dundee DD1 4HN

C. Cowley
School of Psychology,
University of Wales College of Cardiff,
P.O. Box 901 Cardiff CF1 3YG

R. I. Damper
Department of Electronics and Computer Science
University of Southampton
Southampton SO9 5NH

R. Dutton
Centre for Speech Technology Research
University of Edinburgh
80 South Bridge, Edinburgh

J. C. Foster
Centre for Speech Technology Research
University of Edinburgh
80 South Bridge, Edinburgh

C. F. Frankish
Department of Psychology
University of Bristol
8 Woodland Road, Bristol BS8 1TN

K. Hapeshi
School of Psychology
University of the West of England
Bristol

M. G. Helander
Department of Industrial Engineering
State University of New York at Buffalo
Buffalo, NY 14260

D. M. Jones
School of Psychology
University of Wales College of Cardiff
P.O. Box 901, Cardiff CF1 3YG

E. Lewis
Centre for Communications Research
Department of Computer Science
University of Bristol

S. Love
Centre for Speech Technology Research
University of Edinburgh
80 South Bridge, Edinburgh

A. C. Murray
School of Psychology
University of Wales College of Cardiff
P.O. Box 901, Cardiff CF1 3YG

I. R. Murray
The Microcomputer Centre
Department of Mathematics
and Computer Science,
The University
Dundee DD1 4HN

I. A. Nairn,
Centre for Speech Technology Research
University of Edinburgh
80 South Bridge, Edinburgh

A. F. Newell
The Microcomputer Centre
Deparment of Mathematics and
Computer Science, The University
Dundee DD1 4HN

J. M. Noyes
Department of Psychology
University of Bristol
8 Woodland Road, Bristol BS8 1TN

S. Pratt
Department of Communication and
Neuroscience, Keele University
Keele, Staffs. ST5 5BG

F. Stentiford
British Telecom. Research Laboratories
Martlesham Heath
Ipswich, Suffolk IP5 7RE

N. A. Stanton
Department of Psychology
University of Southampton
Highfield, Southampton SO9 5NH

A. F. Creswell Starr
Smiths Industries Aerospace and
Defence Systems, Bishops Cleeve
Cheltenham, Glos. GL52 4SF

M. Tate
Human Factors Research Division
British Telecom. Research Laboratories
Martlesham Heath, Ipswich,
Suffolk IP5 7RE

M. Tatham
Advanced Speech Technology
Laboratory, Centre for Cognitive Science
Essex University
Wivenhoe Park, Colchester, Essex

P. Tucker
School of Psychology
University of Wales College of Cardiff
P.O. Box 901, Cardiff CF1 3YG

D. M. Usher
InterAction, 19 Northampton Street
Bath, Avon BA1 2SN

N. Vergeynst
Centre for Speech Technology Research
University of Edinburgh
80 South Bridge, Edinburgh

R. Webster
Human Factors Research Division
British Telecom. Research Laboratories
Martlesham Heath, Ipswich,
Suffolk IP5 7RE

R. Weeks
Human Factors Research Division
British Telecom. Research Laboratories
Martlesham Heath, Ipswich
Suffolk IP5 7RE

Foreword

M. G. Helander

This text deals with two important technologies in human–computer interaction: computer generation of synthetic speech and computer recognition of human speech. These technologies are quite different and the ergonomics problems in implementation are also different. Nonetheless, synthetic speech and speech recognition are usually dealt with in the same context of speech technology.

Speech technology provides what many think of as the most 'natural' and most efficient mode of human–computer interaction: talk to a computer and get a spoken response back. Indeed, Chapanis and his co-workers confirmed 15 years ago that speech is more efficient than other modes of interaction (Chapanis *et al.*, 1977). However, speech is still difficult to use with computers. In particular, computer speech recognition has not yet been perfected, and to cope with imperfections, users may have to perform unnatural tasks, such as leaving long pauses between spoken words. The question then arises if maybe 'natural' keying is better than 'unnatural' speaking. The other problem has been the limitation in number of recognized words that a computer can understand (its library, or lexicon, of words). This limitation may be less severe than originally thought, since it has been demonstrated that humans can effectively solve problems using a limited vocabulary of about 200 words. In the past few years speech technology has rapidly improved, and we may now be on the verge of solving some of the major technical problems. Such development would have significant economic implications since many viable applications could be implemented. Below, we will briefly review some of the issues addressed in this book.

Synthetic speech

Although advances in speech recognition and natural language understanding have been impressive, neither technology approaches the maturity of speech synthesis. We have today the ability to produce acceptable and intelligible speech from unrestricted text. This doesn't mean that there is no room for improvement, quite the contrary; no one would mistake the best synthetic speech for being generated by a human being, and it is doubtful if anybody would actually prefer to listen to a synthesizer rather than a skilled human reader. Yet among the components of a truly intelligent interface, speech output is clearly what is done best at this time.

Speech synthesizers can of course be justified without the assumption that there is two-way human–computer interaction. They can be used as auditory displays to present varying messages. In the beginning of the 1980s, there were inexpensive speech synthesis chips used in automobiles to warn drivers. The consumer perceived them as annoying and not very informative, and they lasted only two model years. But there are successful applications for the disabled, such as reading machines for the blind and voice output for the vocally handicapped. In addition, this text gives examples of speech synthesis used in avionics.

The improvement of the quality of speech is now the major concern, and several articles in this volume deal with this issue. Commercial synthesizers which generate speech-based on phonemes will inevitably mispronounce words since there are many exceptions to the general rules for pronunciation. To cope with variances in rules, such as 'cough', 'rough' and 'bough', exceptions must be stored in computer memory. Another difficult challenge is to produce the appropriate intonation and emphasis of words in a sentence, referred to as prosodic information. This is helpful to understanding synthetic speech but may be less critical than mispronunciations.

To improve the quality of synthetic speech a new technology is under development. This entails top-down analysis of a sentence, rather than the bottom-up approach used for stringing together phonemes. Individual words are stored in computer memory and sentences are parsed to generate more natural prosodics. Parsing is considered essential, based on the observation that a person reading aloud cannot generate good intonation unless he understands the contents of the text. Even with parsing, it remains difficult to produce appropriate information, but there may be simplifying rules for generating acceptable intonation. After listening to simultaneous translation, I personally get the impression that interpreters cannot possibly internalize everything that is translated, yet their spoken language usually sounds better than the original speaker. According to Streeter (1988) the pronunciation of surnames is the greatest challenge, particularly in the United States where there are 1.5 million uniquely spelled surnames. Even with decreasing costs of computer memory there may be some limitations in the top-down approach.

Additional research topics dealt with in this book include the problems in generating speech with varying precision of articulation and how to convey moods and attitudes. Some of these issues may be less important for the design of usable interfaces, but they are essential for modelling human speech.

Speech recognition

Systems for speech recognition have proven difficult to evaluate in research. The main problem is that the technology keeps improving, and for researchers to make credible comparisons with other input techniques, it is necessary to make projections of where the technology is going. One way to make comparisons is to simulate the best possible recognizer and compare that with the best possible keyboard, or graphics tablet, or joystick. This is the philosophy of the 'Wizard of Oz' scenario where perfect recognition is simulated by a typist hidden to the test subject. Otherwise, ergonomics research has largely been controlled by the algorithms used for recognition, since the idiosyncrasies of the algorithms are the roots of misrecognitions. In this research, error recovery is a major focus and there are many issues related to interface design:

- how can the computer best inform the user about what it understood the user to have said?
- should there be immediate feedback, or should feedback be delayed for a sentence?
- after substitution errors have been discovered, how should they be corrected?
- will the computer pick a good second choice, or should the user try to pronounce the word again?
- should there be different procedures for different error types (false insertion, deletion, rejection)?

To a large extent the answers to these questions depend on the sophistication of the technology.

Another difficulty in research is that users may take several months to learn how to speak appropriately to a computer. This time span is beyond what is considered economical in laboratory studies, and few researchers have had the ability to use real world subjects for longitudinal studies. As a result there are very few long term evaluations available.

One of the major advantages of speech recognition is the ability of the user to perform dual tasks, particularly in conjunction with tasks where the hands and/or the eyes are already occupied. There have been several studies of voice command/control in military environments, and in this text there is evidence that voice recognition is now feasible even in the cockpit. This environment has been considered particularly demanding due to stress-invoked changes in speech patterns and the disturbance from ambient noise.

In this context there are opportunities to perform basic research and apply theories of dual-task performance. For example, according to Wickens' multiple resource theory, verbal tasks are best served by auditory input and speech responses, whereas spatial tasks are best served by spatial (visual) input and spatial (manual) responses (Helander *et al.*, 1988). Would it be possible to validate this theory and propose design guidelines for human–computer interaction?

In the close future we will see new applications for speech technololgy, particularly in multimedia. Maybe this will inspire broader acceptance and use. The time for speech technology may finally have arrived.

References

Chapanis, A., Parrish, R.N., Ochsman, R.B. and Weeks, G.D., 1977, Studies in interaction communication: II. the effects of four communication modes on the linguistic performance of teams during cooperative problem solving, *Human Factors*, **19**, 101–26.

Helander, M.G., Moody, T.S. and Joost, M.G., 1988, Systems design for automated speech recognition, in *Handbook of Human-Computer Interaction* (ed. M.G. Helander) Amsterdam: Elsevier.

Streeter, L.A., 1988, Applying speech synthesis to user interfaces, in *Handbook of Human-Computer Interaction*, M.G. Helander (ed.) Amsterdam: Elsevier.

1

Developing interactive speech technology

C. Baber

Abstract

We are aware that the technical capabilities of speech technology have been developing apace for decades, and that it is on the verge of widespread application. Unfortunately it has been 'on the verge of widespread application' since its inception–what has held it back? One of the key problems the technology faces is the difficulty people have when they first try to use it. This suggests that the 'speech based interaction' between human and computer needs to be carefully designed in order to exploit the strengths of the technology and hide its weaknesses. In this chapter, I will consider the nature of this interaction, drawing on a wide range of disciplines. The aim is to raise, rather than answer, questions concerning this topic.

Introduction

By interactive speech technology, we mean, quite simply, that collection of technologies which allow humans and computers to communicate information etc. through the medium of speech. The word 'computer' is employed here as a generic term for all manner of advanced technology. There has been a great deal of research into the topic, and a number of applications exist in industry (Noyes *et al.*, 1992), in offices (Noyes and Frankish, 1989), telecommunications (Waterworth, 1984a), and in military applications (Chambers and Haan, 1985; Moore, 1989). Furthermore, recent advances in low cost speech technology devices suggest that speech options for personal computers will soon be commonplace. Speech can provide a means of interacting with computers over the telephone, e.g. for travel information, or for 'teleshopping'; it can provide a means for people who ordinarily have difficulty using computers to access and use them (Noyes *et al.*, 1989). However, despite being perennially heralded as the technology whose time has come, the long awaited, widespread application of speech technology has yet to materialize. One of the aims for collecting the papers together in this book is to consider why this should be the case.

> Some speech researchers behave and talk as if they believed that the only bars to widespread adoption of speech interfaces are the intelligibility of speech synthesis, the recognition accuracy and the vocabulary size of speech input devices, and the performance of natural language understanding systems. Only occasionally is any consideration given to dialogue design and other aspects necessary for an effective and efficient human interface...
>
> Newell, 1992

This quotation from Professor Alan Newell of Dundee University encapsulates the ideas behind this book and sets the tone for the papers it contains. Many of the papers first appeared at a one day conference, organized in collaboration with the UK Ergonomics Society, held at the NEC in Birmingham. Researchers from across the UK gathered to discuss those '...other aspects necessary for an effective and efficient human interface...' in the design and development of speech technology. In this chapter, by way of introduction to the main theme of this book, the term 'interaction' will be considered: what do we mean when we speak of an 'interaction' between human and computer, specifically through the medium of speech? While it is currently difficult to provide a definitive answer, by raising and considering this question, it is possible to develop a conceptual framework within which to consider the papers collected in this book. The authors of these papers come from a range of academic and industrial backgrounds, with representatives from the fields of avionics, computer science, ergonomics, linguistics, psychology and telecommunications. However, in many ways the authors share the same underlying goals:

- to seek to understand what can make speech technology function to its best abilities;
- to determine when to use speech technology and, of equal importance, when not to use it; and
- how to design sound, robust and durable speech technology applications.

Readers who may not be familiar with the various technologies considered are advised to consult either Ainsworth (1988a) or Bristow (1986) for an introduction to automatic speech recognition, and Bristow (1984) or Witten (1982) for an introduction to speech synthesis. It is interesting to note the relative dates of these books – the synthesis books are older – and this reflects a common belief that many of the 'problems' of speech synthesis have been dealt with. However, as Tatham (1993) argues, this belief is clearly unfounded, and there is still much useful work which can be conducted in the area of speech synthesis. It is also worth noting that there are currently very few books which specifically address issues of 'speech-based interaction' between human and computer, although Waterworth and Talbot (1987), Taylor *et al.* (1989) and Luff *et al.* (1990) include discussions of some of the issues involved.

Speech as interaction

There have been a number of books published during the 1980s which provide guidelines to designers involved in human computer interaction (HCI). These guidelines range from the voluminous (the HCI guidelines of Smith and Mosier,

1982) to the homely (the '30 proverbs' of Gaines and Shaw, 1984) to the pithy (the '8 golden rules' of Shneiderman, 1987; 1992, see Table 1.1). While these guidelines can often provide useful reference points for the design and evaluation of human computer interfaces (and can be used to untangle some of the ambiguities inherent in the design process) it is difficult to see how they can be used to define specifications for the design of future computer systems, and even harder to see how they can be sensibly applied to the design of speech systems.

For example, Shneiderman's '8 golden rules' were proposed as pointers for 'good' design of any human computer interface. It is worth pausing to ask whether these 'rules' be applied to speech technology, and, if they can, how ought they be applied?

These 'rules' suggest that designers should be striving to produce human computer interfaces which are coherent, i.e. in which similar user actions can be used to produce similar system responses throughout the interaction, and transparent, i.e. in which the appropriate user actions, at any point in the interaction, are clear and obvious to the user. The 'rules' suggest that the user should be permitted to maintain control over the computer, and that the interaction will utilize well defined task sequences. While one can see the relevance of these 'rules' to any form of human computer interaction, problems arise when one attempts to define these 'rules' in such a way as to permit good design of interactive speech technology. As designers, developers and researchers into the field of speech technology, we need to ask what makes speech-based interaction between human and computer coherent or transparent?; should the human always retain control over the interaction?; what are 'well defined task sequences' for speech-based interactions? Only by asking these questions, can we begin to comprehend and define the nature of interactive speech technology.

Speech-based interaction with humans and computers

It is almost inevitable that people will approach speech technology with preconceptions as to how it will work; preconceptions derived, in part, from representations of speech technology in science fiction and, in part, from their everyday experience of speech use in human – human communication. It is this latter domain which most often leads to the unwarranted and naive claims concerning the 'naturalness' of speech as a medium for human computer interaction (see Damper, 1993 for further discussion of this point). However, it is not always easy to overcome the initial expectations

Table 1.1 Eight golden rules for dialog design

- Strive for consistency
- Enable frequent users to use shortcuts
- Offer informative feedback
- Design dialogues to yield closure
- Offer simple error handling
- Permit easy reversal of acts
- Support internal locus of control
- Reduce short-term memory load

[Shneiderman, 1987; 1992]

of prospective users, that speech technology will, or ought to, use speech in a manner which is similar to human - human communication. One complaint we have heard for rejection of speech technology is that it does not appear to be very 'human', either because a synthesized voice sounded too robotic or because a speech recognition device would not deal with certain words.

Somewhat paradoxically, whilst users' expectations of, and attitudes towards, speech technology appear to relate to their experience of speech in human - human communication, their actual use of the technology is noticeably different to human - human communication. Thus, in order to define speech-based interactions with computers, it is necessary to distinguish it from other forms of speech use. That is, in order to appreciate how people speak to computers, we need to consider speech-based interactions as a specific form of speech use. This would suggest that it has certain characteristics which both relate it to, and separate it from, other forms of speech use.

When we first learn to speak, we learn not only the words in a language but also how to combine them into meaningful phrases and how to use them in different contexts. Indeed, babies learn to organize and classify 'vocalizations' before they begin to master the rudiments of speech. A common feature of human - human communication is the use of a 'sequence of speakers.' (Longacre, 1983). If all parties were to speak at once, or if all parties waited for each other to speak, there would be little chance of effective verbal communication. There appear to be strategies which people employ to define who will speak next. Such strategies have been termed 'turn taking' (Sacks *et al.*, 1974), and draw on a wide range of linguistic and extralinguistic cues (Duncan, 1972; 1973; 1974). Thus, we can offer a first pass definition of interaction as a process by which meaningful information is communicated between parties, using a sequence of speakers alternating turns. In order for this process to be effective there needs to be a speaker and an auditor, some set of cues pertaining to turn taking, and some 'understanding' of the language used by both speaker and auditor.

This latter point may present difficulties if we apply the word 'understand', in a conventional sense, to speech-based interaction with computers; clearly a computer is not capable of the breadth and richness of the language understanding abilities of humans. However, as Grosz and Sidner (1986) note, in asking an auditor to perform an action, a speaker need not make any assumption concerning the mental representation by the auditor of the words, only to believe that the auditor is capable of performing that action and that the words used will cue that action. The subsequent performance of the (correct) action would then constitute an acceptable form of 'understanding'. I do not want to proceed further with what could easily become a very thorny philosophical problem, but would like to draw the reader's attention to one potential problem for automatic speech recognition (ASR), based on this argument - how can one compensate for, or otherwise deal with, misunderstanding if the only proof of understanding is the performance of an action? This question, I think, cuts straight to the heart of one of the key issues of interactive speech technology: error correction, which is discussed in several papers in this volume.

Returning to our first pass definition of interaction above, it is possible that users might expect the 'structure' of their interaction to resemble that of human–human communication. Indeed, Murray *et al.* (1993) show that presenting the same information for the same task in different formats, i.e. visual or auditory feedback, can result in changes in performance by the subject. In both conditions, subjects had to say 'no' if a letter was misrecognized. In one condition, subjects were given visual feedback, and in the other they received auditory feedback. While subjects correctly said 'no' in the auditory feedback condition, they often simply repeated the mistaken word in the visual condition. This led Murray *et al.* (1993) to speculate that the auditory feedback may have presented some cue concerning the nature of the turn-taking required in this interaction.

In order to consider this point in more detail, I have produced a 'simulated transcript' of the interactions required in this study. Here we can see two quite distinct types of interaction relating to the different feedback modes.

We can see from this that the interaction in the auditory feedback mode follows a clear progression from initial input to problem to resolution. This progression is emphasized by the use of what Murray *et al.* (1993) term simple intonation contours in the computer speech, with a rising pitch to indicate a question. I believe that this study was conducted using a Votan VCR 2000, in which case the 'computer speech' was digitized from samples provided by the experimenters and replayed at programmed points in the interaction. In the auditory feedback condition, the interaction proceeds by a computer question followed by a user response. In the visual feedback mode, on the other hand, there are no such cues from the computer. The user speaks a letter and monitors the feedback on the screen. Error correction, in this instance, appears to be less a matter of 'dialogue' than of 'editing', and so the user may attempt to repeat the letter to overwrite the one currently displayed. This action, however, violates the interaction requirement for the user to confirm the computer feedback using yes/no. Thus, in one condition the interaction is based on a question/answer routine, but in the other it is based on a command/action routine. This points to the simple conclusion that different interactions will be governed by different turn-taking rules, and that the rules will be prompted by different cues.

Table 1.2 Adjacency pairings in two interaction styles

	Auditory feedback mode		Visual feedback mode
1	U:"P"	1	U:"P"
2	C:"D?"	2	C:<D>
3	U:"no"	3	U:"No"
4	C:"Try Again"	4	C:---
5	U:"P"	5	U:"P"
6	C:"P?"	6	C:<P>
7	U:"Yes"	7	U:"Yes"

Key to Table 1.2
U =user;
C =computer;
" " =spoken information, by either U or C,
< > =visually displayed information,
--- =no response

In any interaction then, there will be legal points of transfer between preceding and succeeding turns. Sacks *et al.* (1974) emphasize this by coining the term **adjacency pair**. At their simplest level, adjacency pairs are interactions which involve each party in a single turn. This notion fits into a view of interaction as a sequence of events in time, and shows how the preceding discourse constrains, and can be used to predict, the development of the interaction.

If we refer back to Table 1.1, we notice that one of the '8 golden rules' is to 'support internal locus of control', i.e. put the user in charge of the interaction. From the notion of turn-taking, we can ask, what are the cues which signal when a change in turn will occur? Human–human communication appears to be very permissive, with a rich array of turn-taking cues, often used in a process of negotiating who will speak next. Human computer interaction, on the other hand, is very coercive requiring one partner to determine who can speak, and using an impoverished set of cues, e.g. an auditory or visual prompt to the user, a command from the user, or some form of feedback from the computer. However, as we have seen, there will be a range of interaction styles available for speech-based interaction with machines and we are unclear as to how different types of cue affect the interaction, and which type of cue would be most suitable for different types of interaction.

Relevance

At this point, we can ask whether it is possible to provide some form of general guidelines for speech-based interaction. Grice (1975) defined a set of maxims for human–human communication, under the general rubric of a **co-operative principle** (Table 1.3). He argued that in order for conversation to function effectively it was necessary for speakers to obey these maxims.

Pinsky (1983) has suggested that it might be possible to employ these maxims in the design of human computer interfaces. For instance, there may be situations when a computer provides information which does not make sense to the user, either

Table 1.3 Maxims derived from Grice's (1975) co-operative principle

Quantity
- Make your contribution as informative as required (for the current purposes of the exchange)
- Do not make your contribution more informative than is required

Quality
- Do not say what you believe to be false
- Do not say that for which you lack adequate evidence

Relation
- Be relevant

Manner
- Be perspicuous
- Avoid obscurity of expression
- Avoid ambiguity
- Be brief (avoid unnecessary prolixity)
- Be orderly

because it is in code provided for the software engineers and interface developers and cannot be interpreted by the user, or because it does not fit in with the user's expectations of what is happening. This would violate the maxims of both quantity and relation. Furthermore, currently operational interactive speech services occasionally suffer from all manner of 'unnecessary prolixity' in their manner of communication, with highly verbose speech messages which seem primarily designed to confuse and overwhelm the user. Table 1.4 gives an example of such speech.

Table 1.4 Initial instructions for an operational interactive speechsystem

"Good afternoon and welcome to **** flight information service. The system was last updated at 20:28 and the last flight time recorded is 22:48. First of all I need to know what type of phone you have. If you have a star button, please press it at this point. If you do not have a star button there will be a short pause."

This example is taken from an operational interactive speech system, which can be accessed from any public telephone. The only alteration made is to substitute **** for the system's name, otherwise this is a verbatim transcription of the speech received when the system was called. As we can see, even in this short sample, the message managed to violate the maxims quantity (by being more informative than is required), relation (by including information which is irrelevant to the caller), and manner (by managing to be, at times, obscure, ambiguous and unnecessarily prolix). There is also at least one point of potential confusion in this message; the user is not informed what function the star button serves, nor what will happen if s/he presses it. One could propose that shortening and simplifying the content of this message would improve its usability. From this example, we can see how Grice's (1975) 'co-operative principle' can be used as the basis for evaluating interactions. However, it is difficult to see how it can be used for the purposes of developing design guidelines.

In an interesting development of these ideas, Sperber and Wilson (1986) collapse the four maxims of the 'co-operative principle' into a single principle of relevance. Briefly, the principle of relevance states that a speaker should try to express information which is most relevant to the auditor. This suggests that, on hearing an utterance, the auditor will make inferences relating the content of the utterance to the situation in which it is uttered. Sperber and Wilson (1986) further suggest that people will employ as little cognitive effort as possible in maintaining relevance. In other words,

> If the utterance is not sufficiently relevant in the initial context, it is unlikely that its **relevance** will be increased by further extensions of the context. Thus, the search for context will cease and the communication will fail.
>
> Stevenson, 1993

This suggests that the spoken interactions people can have with computers will be shaped by a range of factors influencing the definition of context in which the interaction is taking place. If this is the case, then it will be possible to develop means of defining the context of interactions either prior to, or in conjunction with, the design of a vocabulary. Several systems have been developed using such an approach. What is required then, is some definition of the context of an interaction. The next section considers this point in more detail.

Speech as action

What characterizes speech-based interactions with machines? We can begin to answer this question by considering several concepts, proposed by Austin (1962) and extended by Searle (1969; 1975), which combine to form Speech Act Theory. In this theory, Austin (1962) develops a notion of speech as action; he assumes that there are two primary types of utterance, **constantives** (in which a state of affairs is described), and **performatives** (in which speech is used to do things, i.e. '. . . in which by saying. . . something we are doing something.' Austin, 1962).

It is worth noting that this approach has been criticized because, in many ways, it is primarily abstract and difficult to apply to practical problems of human–human communication (Potter and Wetherell, 1987). Furthermore, its emphasis on 'performatives' can make the interpretation of meaning somewhat problematic; in a sense it supposes that words could mean whatever the speaker intends them to mean. While these are obvious limitations when applied to human–human communication, the problem of 'meaning' does not appear to me to be too worrying in speech-based interaction with computers, for the simple reason that, by and large, the 'meaning' of speech is not an issue, and the problem of application can be circumvented by the observation that this chapter is, of necessity, abstract; if there were more examples of working interactive speech applications, it might be possible to develop methods in line with ethnomethodology and discourse analysis. As such systems are still quite rare, I will content myself with abstractions.

As well as having the possibility of having meanings attached to them, Austin (1962) argues that utterances perform actions according to their 'force', e.g. while the words *close the window* have a specific meaning, they can also be uttered to convey the force of an order or a command or a request or as a means of annoying someone. In other words, the force of an utterance can be described as its extralinguistic meaning in a particular context. From this, Austin (1962) proposed three basic types of speech act: locutionary – in which the sense and reference of an utterance can be clearly determined; illocutionary – in which utterances achieve an effect via their force, e.g. ordering or promising; perlocutionary acts – in which utterances produce effects upon their auditors with respect to the context in which they are uttered. Thus, we have the preliminaries of the pragmatics of discourse. These notions have led to some research in discourse modelling in artificial intelligence (Perrault, 1989), based on the formalized conception of speech acts between human and computer as follows:

> . . .the speaker performs a specific illocutionary act, with the intention of eliciting a perlocutionary effect in his/her auditor, in order to achieve a specific goal. The auditor, in recognizing both the content and the illocutionary force of the speaker's utterance, responds accordingly.
>
> Perrault, 1989

In order to understand the nature of speech-based interaction with computers, then, we need to consider the manner in which illocutionary acts can be performed by users. In this way, we can begin to develop a taxonomy of interaction styles, and examine how different styles can be made effective in different situations. This will permit an extension of the example, taken from the Murray *et al.* paper and used above to consider not only feedback format but also user intentions and goals, choice of appropriate vocabulary etc. Searle (1975) distinguished five types of illocutionary act as given in Table 1.5.

Whether or not one agrees with the systematicity of this taxonomy, one can see points of discrepancy between human–human communication and speech-based interactions with machines. For instance, several studies have shown that, while 'expressives', i.e. formulaic utterances expressing politeness, thanks and apologies, are common in human–human communication, they are virtually non-existent in speech-based interactions performed in the same task domains (Richards and Underwood, 1984; Meunier and Morel, 1987; Baber, 1991). Indeed, Baber and Stammers (1989) found notable differences in utterance type between subjects performing a task, depending on whether or not they believed that they were speaking to a computer. The task simply required subjects to identify boxes on a screen, using a colour and a number. The experiment involved a 'Wizard of Oz' subterfuge, in which an experimenter listened to subjects' speech and activated computer responses using a dedicated, function keyboard. When subjects knew that they were speaking to an experimenter, many of their commands included the words 'please', or 'would you . . .', and, in a sizeable minority of cases, subjects followed 'computer responses' by the phrase 'thank you'. In none of the cases when subjects believed that they were speaking to a computer did these utterance types occur. Thus, there appeared to be distinct 'conversation management protocols' at work in these conditions; subjects speaking to the experimenter retained a degree of social contact in their communication. Indeed, it would appear that the 'social contact' was actually exaggerated by the experimental situation. In the cases where subjects believed that they were speaking to a computer, there did not appear to be a necessity for establishing and maintaining a 'social contact' through the use of expressives.

With respect to the other types of illocutionary act identified by Searle (1975), it is difficult to imagine situations in which either 'representatives' or 'commissives' will have direct relevance to speech-based interactions; the speech technology system will not be able to verify the truthfulness of a statement, only whether or not the system can respond, and, as speech systems are used for immediate task performance, the idea of a user promising to do something at a later date seems a trifle implausible.

Table 1.5 Types of illocutionary act

• Representatives	in which a speaker is committed to the truth of a statement
• Directives	in which the speaker tries to get the auditor to do something
• Commission	in which a psychological state is expressed
• Expressives	in which a psychological state is expressed
• Declarations	in which an utterance is used to change the state of affairs

[after Searle, 1975]

Thus, we are left with 'directives' and 'declarations' as the principal types of illocutionary speech act by which speech-based interaction with machines will be conducted, i.e. acts in which speech can be said to directly modify the state of the auditor, either through making it perform a specific response or providing it with specific items of data.

Speech act theory has been adapted by Bunt *et al.* (1978) and Waterworth (1984b), and Table 1.6 illustrates their approach. In a given interaction, there will be some form of boundary marking for the start and termination of the interaction, such as greetings/farewells, and also for the start and termination of each individual utterance within that interaction, such as prompts from the computer to signal the user to speak, or a signal from the computer to indicate that something has been recognized. Within the interaction, there will be both some form of 'interaction management protocol' to ensure that the interaction is progressing smoothly, and 'goal directed' acts. In Table 1.6, I have focused on the interaction management acts, as the goal directed acts would require an example of a specific interaction scenario.

There seem to be a number of possible uses of interaction management protocols for speech-based interactions. Waterworth (1982) suggests that their primary use is to prevent and correct mistakes. But one could also view their usage as ways of ensuring the Grice's (1975) 'co-operative principle', or Sperber and Wilson's (1986) 'relevance principle', are being adhered to, i.e. to determine whether sufficient

Table 1.6 Speech acts in speech-based interaction with machines

Boundaries	Goal directed	Dialogue acts Interaction management
*start		
	*greeting	
		[*enrolment/speech sampling]
		prologue
		*preamble
		*explanation
	*instructions	
		*vocabulary items
*interaction	speaker→auditor	• grunt
		• yes/no
		• direct
		• command
		• request
		• recovery/repair
	auditor→speaker	• confirm
		• acknowledge
		• accept
		• reply
		• no response
		• recovery/repair
		• provide extra information
*termination	*quit	
	*sign off etc.	

information has been communicated to enable the auditor to understand what has been said, to control the rate at which information is passed, to request or give additional information, or to stop the communication if redundant or repeated information is being passed.

In human–human communication, conversation management protocols can be used to reduce the size of the required vocabulary. This results from the development of a shared terminology, based on common knowledge (Krauss and Glucksberg, 1974; Falzon, 1984), and on the use of short, simple feedback phrases which limit and control the amount and rate of information exchanged (Oviatt and Cohen, 1989). This latter form of control is often performed using paralinguistic devices, known as 'backchannelling', e.g. 'mm-hmm' or 'huh?' (Schoeber and Clark, 1989). However, in order for such acts to function effectively, it is necessary for both parties to have some knowledge of the conventions by which they are communicating. This has led to some writers calling for the introduction of additional knowledge, concerning the use of intonation for instance, in automatic recognition speech systems in order to improve performance (Leiser, 1989a,b). The goal of providing such additional knowledge would be to emulate human communication conventions, but as we have already noted, there is no guarantee that speech-based interaction with machines will adopt the same conventions. What is needed is some means of describing the conventions by which speech-based interactions with machines actually function.

In Table 1.6, there are a number of 'acts' which are placed between the 'goal directed' and 'interaction management' columns; this is to illustrate that there will be several acts which are used to link parts of the interaction. For instance, the 'greeting' will be partly an 'interaction management' act, in that it will serve to initiate the interaction, and it will be partly goal directed in that it will be performed with the twin goals of providing specific introductory information and establishing the appropriate interaction style. This latter point is based on the observation (discussed below), that the spoken feedback from the computer can help to shape both the users' perceptions of, and attitudes towards the computer, and the manner in which the user will speak to the computer. From Table 1.6 it is conceivable that we can begin to define certain stereotypical structures, e.g. greetings/farewells, which can be used as building blocks for the design of interactions. For instance, Table 1.7 shows some of the possible speaker acts and turn types which could be found in different forms of speech-based interaction. The turn types represent four simple adjacency pairs: question and answer (Q–A), data-entry and logging (DE–L), data-entry and entry (DE–E), and command and response (C–R).

For instrumental feedback, the user will be supplied with immediate confirmation of recognition, either in the form of a 'beep' or the machine performing the appropriate action; for operational feedback, the user will be supplied with the words recognized and will need to make a decision as to the validity of the recognition.

From the discussion so far, we have emphasized that speech-based interactions with machines can be considered as action, i.e. intentional, goal-based activity, which is performed within a definable context, and differs from human–human communication in its range of styles. The manner in which such actions will be fitted together will be dependent on a wide range of factors, as given in Table 1.8.

Table 1.7 Possible speaker acts in speech-based interactions

Speaker	Examples	Application	Auditor	Feedback signal	Turn type
"grunt"	hm-mm	simple query system	acknowledge	operational	Q–A
Yes/no	"yes" or "no"	simple menu	select choice	operational	Q–A
direct	1) "London"	baggage handling	accept	instrumental	DE–L
	ii) "2.45"	faceplate inspection	accept	instrumental	DE–L
	iii) "6 in cell 8"	spreadsheet	"6" in cell	instrumental	DE–E
command	i) "open valve 4"	telecommand	valve opens	instrumental	C–R
	ii) "turn on lights"	domestic	light on	instrumental	C–R
	iii) "lift right arm"	robotics	arm moves	instrumental	C–R
request	i) "when does the next train to London leave?"	timetable services	"11.47"	operational	Q–A
	ii) "Output block 8"	process control	"249 volts"	operational	C–R
	iii) "Show product cells 6 and 8"	spreadsheet	"148"	operational	C–R

Table 1.8 Factors which may influence the efficiency of speech-based interaction with computers

(i) Context
- task domain in which the interaction is being performed
- environment
- communication channel, e.g. radio links

(ii) Media
- format used for computer output, e.g. visual or auditory
- quality/intelligibility of computer speech
- recognition accuracy
- interface design

(iii) Language
- 'co-operative' and 'relevance' principles
- similarity between computer and user language
- restrictions placed on users' language, in terms of vocabulary and speech style

(iv) Interaction
- level of uncertainty (see below)
- turn taking
- model of partners employed
- degree of shared knowledge/expertise
- error handling
- ease of learning

(v) User
- level of skill and experience
- response to environmental factors, e.g. stress
- speech characteristics
- individual differences

This list is by no means exhaustive, but it serves to illustrate some of the factors which could contribute to the manner in which a speech-based interaction will be conducted. The following section will consider how such factors serve to shape the interaction.

Shaping interaction?

If we assume that a characteristic of speech-based interaction with machines is its limited channel of communication and vocabulary, we can draw some parallels between it and 'similar' human–human communication activities; these are illustrated by the following examples:

1. human communication over the telephone, lacking the richness of face to face communication, is limited to short phrases and, often, task oriented (Fielding and Hartley, 1987; Rutter, 1987; Jaffe and Feldstein, 1970);
2. in doctor/patient discourse, around 90 per cent of the doctors' turns took the form of questions (Frankel, 1984), i.e. a single type of adjacency pair;
3. when 'experts' and 'novices' co-operate to solve a problem, the dialogue is developed through a negotiation of knowledge (Grosz, 1977; Falzon, 1991), i.e. the partners will reach agreement on how best to describe objects in terms which they can both understand.

Therefore, the relationship between partners, in terms of their relative degrees of expertise and the communication medium used, will shape the dialogue.

There has been some interest recently in whether it is possible to use the feedback from a computer to shape the user's speech. In human communication, people often, and presumably without being aware that they are doing so, imitate mannerisms and forms of expression of the interlocutor. This is a phenomenon known as convergence (Giles and Powisland, 1975), and effects such properties of a human dialogue as pause length, and use of particular linguistic structures (Schenkein, 1980). A number of studies have found that, given unrestricted speech input, subjects would often adapt to the 'linguistic style' of the computer (Zoltan Ford, 1984; Ringle and Halstead Nussloch, 1989; Leisser, 1989a, 1989b), and, in some instances, the intonation of the speech synthesis device. Thus, when users are free to employ whatever commands and vocabulary they thought necessary, they often used terms and structures that the computer used in its responses. This does indeed suggest that it might be possible to use feedback to shape, and constrain, users' speech. However, it is worth asking whether these findings result from the linguistic phenomenon of convergence, or whether there may be another explanation.

Baber (1991) reports studies investigating the use of speech for a simple process control task. In one study, subjects were provided with either 'verbose' or 'limited' feedback, on the assumption that feedback type would influence performance in a manner predicted from the notion of convergence. While feedback type did influence performance, it was actually in the opposite direction to that expected. When feedback length increased, i.e. was verbose, subjects responses became shorter. Meunier and Morel (1987) have noted a similar 'paradox', and it is worth quoting their observation in full:

> The length of the speaker's and simulated computer's utterance is inversely proportional: When the simulated computer produces very long sentences without ellipsis or anaphora, the speaker produces very short sentences, reducing them to the strictly necessary constituents.
>
> Meunier and Morel, 1987

This suggests that users' responses are related to the degree of uncertainty they have of the 'partner's capabilities'. In Baber (1991) it was proposed that the 'verbose' feedback implied some necessity for 'negotiating' the subject's commands, such as providing some explicit confirmation of the recognition. However, such 'negotiation' was both inappropriate to the task and to the interaction, which, in turn, lead to some uncertainty on the part of the subject. In order to reduce this uncertainty, subjects limited their speech to the essential words. In other interactions, such as the one shown in Table 1.4, the speech output from the computer may be inducing a heavy memory load on the user, and as a way of compensating for this, the user may choose to respond with as short a phrase as possible. Both instances related to the central notion of uncertainty reduction; in the first example, there was uncertainty as to the appropriate style of speech, and in the second example, there was uncertainty concerning information load.

This notion of 'uncertainty reduction' also offers, I feel, a more parsimonious explanation of the 'convergence' data. In speaking to computers, people are faced with an ambiguous interaction partner whose linguistic capabilities are not directly inferable. This means that users will need to deal with a high degree of uncertainty as to what constitutes an appropriate form of speaking. In order to reduce this uncertainty, they will try to limit their speech to a style which they believe will '. . . more easily enable machine analysis.' (Richards and Underwood, 1984), or will try to employ elements of the 'machine's' vocabulary into their speech. This latter point suggests that rather than be an unconscious process of 'convergence', the use of the machine's vocabulary is a conscious strategy aimed at reducing uncertainty. This uncertainty will be exacerbated by machine output which violates 'co-operative principle' proposed by Grice (1975). Following the notion of minimal effort (Sperber and Wilson, 1986), we can also assume that users will use those aspects of a vocabulary which are as simple and succinct as possible to achieve specific goals. Thus, their speech will be of sufficient quantity and quality to achieve their goal, whilst matching the speech-processing capabilities which the user attributes to the computer.

When people use speech-based interactions they will be forming impressions of the capabilities of the computer they are speaking to and will try to match their speech to the capabilities of that computer. They will be aiming the utterances towards attainment of some goal, and will use sufficient speech for this aim. They will seek to reduce ambiguity and uncertainty in the interaction by maintaining the initiative, most probably by restricting their own speaking turns to try to control the rate at which turns are taken. They will attempt to reduce short term memory load by restricting their speech. Naturally these hypotheses will not apply to all interactions in all applications, but this is difficult to substantiate due to the lack of work in the area. However, what these hypotheses do suggest is that people are not passive users of speech technology, but 'create' an interaction partner, based on the information presented by the computer and the manner in which the information is presented, and it is this created partner to whom they speak.

We can also note here that, in common with human communication, people will attempt to use speech styles which require the least amount of effort. Clark and

Shaefer (1987), examining error correction strategies in directory enquiry telephone conversations, observed that callers would tend to repeat only the section of a digit string in which a misrecognition occurred. Several studies examining speech-based interaction with machines have found that people will try to correct recognition errors by repeating the intended word, rather than using the required error correction dialogue (Baber *et al.*, 1990; Leggett and Baber, 1992).

Conclusions

From the preceding discussion, I hope it will be apparent that we can characterize some of the aspects of speech based interaction with machines. However, it is worth noting that it is still the case that, despite being able to provide such characterization, it is clear that we lack a rigorous linguistics of human-computer interaction.

If we wish to progress beyond the 'trial and error' design of interactive speech systems it is necessary to develop a systematic theoretical account of how people can be expected to interact with speech technology. In order to do this, we need to consider the use of speech technology for a host of applications.

Traditionally, the design process begins with a specification, and technology is developed to meet that specification. If we accept this, then speech technology already appears to be facing something of a dilemma; the technology already exists and we are in the business of applying it. Peckham (1984) asks whether speech technology can be considered '. . .a gimmick or a solution in search of a problem?'. There are several applications, particularly of speech synthesis, in which poor implementation of low quality products have received a high incidence of press coverage and achieved the status of infamy, e.g. with stories of owners of talking cars being 'driven to distraction', and passengers in talking lifts disabling loudspeakers to silence the 'annoying voice'. It appears that the majority of interactive speech systems built to date have relied very heavily on the intuitions of the designers. This is not to criticize the designers, but to indicate that they had very little option because there is simply insufficient information regarding how people do, and how they ought to, speak to machines. An alternative approach is to use information from human–human communication as the basis for system design. This has the unfortunate consequence of leading to interactions with 'natural' human characteristics which are inappropriate to interaction with machines. A further alternative is based on the notion of 'user-centred design' in which prospective users, or a representative sample, use and evaluate prototypical versions of the product. Although this involves potential users, and can provide some useful insights into problems of usage, this methodology suffers from the problem that users very often do not know what it is that they want from a technology as novel as speech. Furthermore, by using a working system, designers could be in danger of only fine-tuning the existing design rather than ensuring that they have achieved that 'best' design. The final approach is to simulate a speech technology system, and use this simulation to modify and test specific aspects of the interaction. It is this final approach which is being developed and used by a number of researchers (Tate *et al.*, Chapter 18; Foster *et al.*, Chapter 19).

We can use simulated interaction, based on the Wizard of Oz approach (Kelley, 1983) to study people in speech-based interactions with machines; we can use flowcharting techniques to describe the flow of the interaction (Leggett and Baber, 1992), and we can develop mathematical models of dialogue parameters (Ainsworth, 1988b). Of course, it is important to view design as an incremental, rather than absolute, process, but it helps to have the right material to start with and that material ought to be derived from human interaction with machines rather than with other people.

This chapter presents simple descriptions of different types of interaction and shows how different forms of feedback from a speech system can shape the user behaviour. One can extend these points and suggest that different forms of interaction, i.e. enquiry, command etc., will have different interaction requirements. This is, or ought to be self evident, but is not necessarily used as a basis for design. I have sought to raise some of the problems associated with the design and implementation of interactive speech technology, and no doubt, during the process of reading these points, the reader will have generated a few more. There are, of course, still many questions concerning what we mean by interaction, and how it will work, but I hope that this chapter will give some pause for thought as to how such questions can be dealt with. In the other papers in this book, specific aspects of the use, design and evaluation of interactive speech technology are discussed.

References

Ainsworth, W. A., 1988a, *Speech Recognition by Machine*, London: Peter Peregrinus.

Ainsworth, W. A., 1988b, Optimization of string length for spoken digit input with error correction, *International Journal of Man Machine Studies*, **28**, 573–81.

Austin, J. L., 1962, *How to do things with words*, Oxford: OUP.

Baber, C., 1991, *Speech technology in control room systems: a human factors perspective*, Chichester: Ellis Horwood.

Baber, C. and Stammers, R. B., 1989, Is it natural to talk to computers: an experiment using the Wizard of Oz technique, in E. D. Megaw (Ed.), *Contemporary Ergonomics 1989*, London: Taylor & Francis.

Baber, C., Stammers, R. B. and Usher, D. M., 1990, Error correction requirements in automatic speech recognition, in E. J. Lovesey (Ed.), *Contemporary Ergonomics 1990*, London: Taylor & Francis, 454–59.

Bristow, G., 1984, *Electronic Speech Synthesis*, London: Collins.

Bristow, G., 1986, *Electronic Speech Recognition*, London: Collins.

Bunt, H. C., Leopold, F. F., Muller, H. F. and van Katwijk, A.F.V., 1978, In search of pragmatic principles in man–machine dialogues, *IPO Annual Progress Report 13*, 94–8.

Chambers, R. M. and de Haan, H. J., 1985, Applications of automated speech technology to land-based army systems, *Speech Technology* (Feb./March) 92–9.

Clark, H. H. and Schaefer, E. F., 1987, Collaborating on contributions to conversations, *Language and Cognitive Processes*, **2**(1), 19–41.

Damper, R. I., 1993, Speech as an interface, in C. Baber and J. M. Noyes (Eds), *Interactive Speech Technology*, London: Taylor & Francis.

Duncan, S., 1972, Some signals and rules for taking speaking turns in conversation, *Journal of Personality and Social Psychology*, **23**, 282–92.

Duncan, S., 1973, Towards a grammar for dyadic conversation, *Semiotica*, **9** 29–47.

Duncan, S., 1974, On the structure of speaker–auditor interaction during speaking turns, *Language in Society*, **2**, 161–80.

Falzon, P., 1984, The analysis and understanding of an operative language, in B. Shackel (Ed.), *Interact '84*, Amsterdam: Elsevier.

Fielding, G. and Hartley, P., 1987, The telephone: a neglected medium, in A. Cashdan and M. Jordan (Eds), *Studies in Communication*, Oxford: Basil Blackwell.

Frankel, R., 1984, From sentence to sequence: understanding the medical encounter through microinteractional analysis, *Discourse Processes*, **7**, 135–70.

Gaines, B. R. and Shaw, M. L. G., 1984, *The Art of Computer Conversation*, Englewood Cliffs, NJ: Prentice Hall.

Giles, H. and Powisland, P. F., 1975, *Speech Styles and Social Evaluation*, London: Academic Press.

Grice, H. P., 1975, Logic and communication, in P. Cole and J. L. Morgan (Eds), *Syntax and Semantics III: Speech Acts*, NY: Academic Press.

Grosz, B. J., 1977, *The Representation and Use of Focus in Dialogue Understanding*, Menlo Park, CA: Centre Technical Report 151.

Grosz, B. J. and Sidner, C. L., 1986, Attention, intention and the structure of discourse, in A. Joshi, B. Webber and I. Sag (Eds), *Elements of Discourse Understanding*, Cambridge: CUP.

Jaffe, J. and Feldstein, S., 1970, *Rhythms of Dialogue*, London: Academic Press.

Kelley, J. F., 1983, An empirical methodology for writing user-friendly natural language computer applications, *Procededings of CHI '83*, NY: ACM.

Krauss, R. M. and Glucksberg, S., 1977, Social and nonsocial speech, *Scientific American*, **236**(2), 100–05.

Leiser, R. G., 1989a, Exploiting convergence to improve natural language understanding, *Interacting with Computers*, **1**(3), 284–98.

Leiser, R. G., 1989b, Improving natural language speech interfaces by the use of metalinguistic phenomena, *Applied Ergonomics*, **20**(3), 168–73.

Leggett, A. and Baber, C., 1992, Optimising the recognition of digits in automatic speech recognition through the use of 'minimal dialogue', in E. J. Lovesey (Ed.), *Contemporary Ergonomics 1992*, London: Taylor & Francis, 545–50.

Longacre, R. E., 1983, *The Grammar of Discourse*, NY: Plenum Press.

Luff, P., Gilbert, N. and Frohlich, D., 1990, *Computers and Conversation*, London: Academic Press.

Meunier, A. and Morel, M-A., 1987, Les marques de la demande d'information dans un corpus de dialogue homme–machine, *Cahiers de Linguistique Française*, **8**.

Moore, T. J., 1989, Speech technology in the cockpit, in R. S. Jensen (Ed.), *Aviation Psychology*, Aldershot: Gower Technical.

Murray, I. R., Newell, A. F., Arnott, J. L. and Cairns, A. Y., 1993, Listening typewriters in use: some practical studies, in C. Baber and J. M. Noyes (Eds), *Interactive speech technology*, London: Taylor & Francis.

Newell, A. F., 1992, Whither speech systems? Some characteristics of spoken language which may effect the commercial viability of speech technology, in W. A. Ainsworth, (Ed.), *Advances in speech, hearing and language processing*, 2, London: JAI Press Ltd.

Noyes, J. M. and Frankish, C. F., 1989, A review of speech recognition applications in the office, *Behaviour and Information Technology*, **8**(6), 475–86.

Noyes, J. M., Baber, C. and Frankish, C. F., 1992, Automatic speech recognition in industrial applications, *Journal of the American Voice I/O Society*, **18**, 1–25.

Noyes, J. M., Haigh, R. and Starr, A. F., 1989, Automatic speech recognition for disabled people, *Applied Psychology*, **20**(4), 293–8.

Oviatt, S. L. and Cohen, P. R., 1989, The effects of interaction in spoken discourse, Proceedings of the 27th Annual Meeting of the Association for Computational Linguistics, Vancouver, BC.

Peckham, J. B., 1984, Speech recognition–what is it worth?, in J. N. Holmes (Ed.), Proceedings of the 1st International Conference on Speech Technology, Amsterdam: North Holland.
Perrault, C. R., 1989, Speech acts in multimodel dialogue, in M. M. Taylor, F. Neél and D. G. Bouwhis (Eds), *The Structure of Multimodal Dialogue*, Amsterdam: North Holland.
Pinsky, L., 1983, What kind of 'dialogue' is it when working with a computer?, in Green, T. R. G., Payne, S. J., Morrison, D. L. and Shaw, A. (Eds), *The Psychology of Computer Use*, NY: Academic Press.
Potter, J. and Wetherell, M., 1987, *Discourse and Social Psychology*, London: Sage.
Richards, M. A. and Underwood, K. M., 1984, Talking to machines: how are people naturally inclined to speak?, in E. D. Megaw (Ed.), *Contemporary Ergonomics 1984*, London: Taylor & Francis.
Ringle, M. D. and Halstead Nussloch, R., 1989, Shaping user input: a strategy for natural language dialogue design, *Interacting with Computers*, **1**(3), 227–44.
Rutter, D. R., 1987, *Communication by Telephone*, Oxford: Pergamon Press.
Sacks, H., Schlegoff, E. and Jefferson, G., 1974, A simple systematics for the organisation of turn taking for conversation, in J. Schenkin (Ed.), *Studies in the Organisation of Conversational Interaction*, NY: Academic Press.
Schenkein, J., 1980, A taxonomy of repeating action sequences in natural conversation, in B. Butterworth (Ed.), *Language Production I*, NY: Academic Press.
Schoeber, M. F. and Clark, H. H., 1989, Understanding by addressees and overhearers, *Cognitive Psychology*, **21**, 211–32.
Searle, J. R., 1969, *Speech Acts*, Cambridge: CUP.
Searle, J. R., 1975, A taxonomy of illocutionary acts, in K. Gunderson (Ed.), *Minnesota Studies in the Philosophy of Language*, Minnesota: University of Minnesota Press.
Shneiderman, B., 1987, *Designing the User Interface*, Reading, Mass: Addison-Wesley.
Shneiderman, B., 1992, *Designing the User Interface*, 2nd Edn, Reading, Mass: Addison-Wesley.
Smith, S. L. and Mosier, J. N., 1982, *Design Guidelines for User-System Interface Software*, Bedford, MA: MITRE Corp. Report ESD-TR-84-190.
Sperber, D. and Wilson, D., 1986, *Relevance: Communication and Cognition*, Oxford: Basil Blackwell.
Stevenson, R. J., 1993, *Language, Thought and Representation*, Chichester: Wiley.
Taylor, M. M., Neél, F. and Bouwhis, D. G., 1989, *The Structure of Multimodal Dialogue*, Amsterdam: North Holland.
Tatham, M., 1993, Voice output for human-machine interaction, in C. Baber and J. M. Noyes (Eds), *Interactive Speech Technology*, London: Taylor & Francis.
Waterworth, J. A., 1982, Man-machine speech dialogue acts, *Applied Ergonomics*, **13**, 203–7.
Waterworth, J. A., 1984a, Speech communication: how to use it, in A. Monk (Ed.), *Fundamentals of Human Computer Interaction*, London: Academic Press.
Waterworth, J. A., 1984b, Interaction with machines by voice: a telecommunications perspective, *Behaviour and Information Technology*, **3**, 163–77.
Waterworth, J. A. and Talbot, ?., 1987, *Speech and Language based Interaction with Machine: towards the conversational computer*, Chichester: Ellis Horwood.
Witten, I. H., 1982, *Principles of Computer Speech*, London: Academic Press.
Zoltan Ford, E., 1984, Reducing the variability in natural language interactions with computers, Proceedings of the 28th Annual Meeting of the Human Factors Society, Santa Monica, CA: Human Factors Society, 768–72.

Part I: Speech Output

2

Speech output

C. Baber

Introduction

Science fiction has long made the notion of a talking computer highly familiar to us. The application of speech output in such products as cars and lifts or to such services as telephone directory enquiries, has reinforced this familiarity. We may even feel that the problems of speech output have now been solved, and that it will only be a short time before all manner of gadgets, from automated telling machines to electric toasters, begin talking to us with that slightly condescending, robotic voice that speech output tends to favour. The infamous case of a 'talking car', noted by Martin Helander in the Foreword to this volume, urges some caution in the rush to apply synthetic speech.

Speech acts for speech output

Before discussing the papers in this section, I will consider speech output from the perspective of interactive speech technology, i.e., what are the 'speech acts' for which speech output can be used? In very broad terms, speech output will either be supplied in response to a user request, or be supplied independently of a user request. The latter class of acts includes warnings, advisories, commands, prompts and instructions. Each of the various acts will have defining characteristics which makes their requirements somewhat different to other acts. This means that it makes sense to consider speech output from the requirements of specific applications, rather than in general terms.

Speech output in response to user input

Using speech output to respond to user queries

There are many applications, both real and proposed, in which speech output from a computer is used to provide information to system users, often over telephone networks. Application areas range from travel information to banking services.

With reference to the latter application, the National Bank of North Carolina has recently installed a speech output system to handle customer enquiries. The bank handles an estimated 2 million calls per year, with around 90 telephonists. The telephonists had perceived the job as stressful, with the high volume of calls, and mundane, with the sheer repetitiveness of calls. Installation of an interactive speech system, allowing speaker independent, isolated word recognition to access speech output, provided customers with a fast response to routine queries, such as account balance, while leaving the telephonists free to deal with more complex queries (Crosbie, 1991). The cost of installing the system had been recouped within the first three months of application; customers were charged 25 cents per call (in the previous system, calls were charged at 50 cents). There was a 44 per cent increase in the number of computer answered customer calls in the first two years of installation.

More recent applications include the use of speech output in automated telling machines (ATMs), also known as cashpoints, where subjective responses to the use of speech output has tended to be unfavourable. The public nature of speech output often works against it; people are not keen for their account balance to be broadcast to other ATM users.

In these applications, it is important for the speech to be intelligible to the user, especially if the information is being transmitted over a noisy telephone line. If the speech is difficult to understand, it is unlikely that people will wish to continue using the technology. Furthermore, the amount of information that people can attend to and remember is limited when the information is presented in speech.

Feedback from control inputs

A number of applications use speech output as a means of echoing users' input, either directly by simply repeating the words used, or more subtly, as in 'did you say twenty pounds?' Here, the question is used both to echo the user's input and also to confirm the devices' recognition. There are, of course, limits to where such echoing is useful, and where it becomes either intrusive or simply irritating.

Speech output independent of user input

Warnings

Speech output can be very intrusive, and can interrupt current task activity. In terms of warnings, the attention grabbing aspect of speech output can be very useful. In this case speech output provides an 'eyes free' information display, which allows users to receive information while attending to other displays or tasks. Furthermore, the speech output can be displayed to a number of people, i.e., it can be public, which may be a useful aspect for system types of warning display. In this instance, it is often advisable to make the speech output stand out from background speech. Some writers advise the use of a 'machine voice', which sounds distinct from human speech. Obviously, this should not be at the expense of losing intelligibility.

Advisories

Advisories are warnings of potential problems, such as 'advise fuel low'. One of the principle problems arising from the use of advisories is a tendency to 'cry wolf'. A potential problem may either right itself or not be of interest to the user. If too many irrelevant advisory messages are presented there may be a tendency for users to disable the speech output system.

Commands

Some research has attempted to use speech output to issue commands. This is perceived to be useful in proceduralized activity, telling a user to 'perform task X now'. However, many users report resentment at being told to do something by a machine.

Prompts and instructions

An increasingly popular use of speech output is in providing instructions for users of sophisticated technology, such as video cassette recorders or ATMs. It has proved a popular application in toys, and can provide information on how to use a product.

Material covered in section I

This section contains three chapters which approach the issue of speech output from very different perspectives.

Chapter 3 contains both a concise review of the various technologies associated with speech output and a critique of the complacency surrounding its development. Marcel Tatham has been active in the field of linguistics for many years, and brings his experience to the development of a novel approach to generating speech output, in the SPRUCE system. He argues that, while linguists have been interested in the building blocks of speech, contemporary linguistic theories of how speech is put together has little to offer developers of speech output devices – there is no reason why they should; we do not ask researchers from other human sciences to provide theories which are applicable to engineering problems. Paradoxically, the use of linguistic theory in the development of speech output, then, is somewhat misguided. Tatham argues for a new underlying theory – cognitive phonetic theory. Combining cognitive phonetic theory with detailed models of the pragmatic effects of speech output, using a neural network, has led to very high quality speech output. The level of quality achieved ought to pave the way for a new generation of speech output applications.

Chapter 4 considers the application of speech output in the field of computer-aided learning. Eric Lewis begins by arguing that it is strange, considering the ubiquity of speech in conventional teaching practices, that computer-aided learning (CAL) does not make more use of speech output. Previous attempts at introducing speech into CAL tended to rely on taped speech, which was both cumbersome and difficult to use; very often the spoken message was being played at the wrong point in an instruction session. The development of good quality, cheap synthesized speech output means that it is possible to introduce speech into CAL. Referring to the speech acts outlined above, one can imagine speech output in CAL being used principally for prompts, instructions and responses to queries, but also for all the other acts, depending on the application. The use of speech output for CAL would seem to be useful proving ground for technical developments.

Chapter 5 presents a less optimistic view, and considers how useful speech output could be in a process control task. Neville Stanton reports an experimental comparison of speech output with a text display for alarm information, i.e., warnings. His results show that, contrary to expectations, speech displays do not necessarily yield faster, more accurate response than visual displays. In this experiment, not only did the speech output produce slower performance than a text display, but also produced more inappropriate activity. There are a number of possible explanations for this finding, e.g., subjects in the speech condition waited until the complete speech alarm was produced before responding (Ito *et al.*, 1987); the quality of the speech output was quite low, although it was produced by a commercially available synthesis chip; speech did not 'map' well onto subjects' conceptions of the task (Robinson and Eberts, 1987). These points are considered in more detail by Stanton. The main point to note is that care must be taken before the application of speech output to a task domain.

Taken together, the chapters in this section provide an overview of research trends in speech output: efforts are being made to define appropriate application domains, experimental comparisons are being made between speech and other media, and research is underway to produce high quality speech output.

References

Crosbie, D., 1991, Voice response systems increase customer satisfaction for a large USA bank, *Speech Technology*, **5**, 450–61.

Ito, N., Inoue, S., Okura, M. and Masada, W., 1987, The effect of voice message length on the interactive computer system, in F. Klix, A. Streitz, Y. Waern and H. Wandke (Eds), *Man-Computer Interaction Research: Macinter II*, Amsterdam: North Holland, 245–52.

Robinson, C. R. and Eberts, R. E., 1987, Comparison of speech and pictorial displays in a cockpit environment, *Human Factors*, **29**, 31–44.

3

Voice output for man–machine interaction

Marcel Tatham

Abstract

Text-to-speech synthesis has been available for use with interactive applications for thirty years, but only now are systems beginning to emerge which will meet the requirements of versatility of usage and naturalness of output. There have been two major reasons for the delay: inadequate attention to the underlying theory of speech production in human beings, and the widespread belief that the difficulties had already been solved. This paper explains the theoretical short-comings of current systems of text-to-speech synthesis, and describes one major system, SPRUCE, which addresses the updating of the speech production model directly. It is argued that the time is right for greater interest and investment in voice output systems for interactive applications.

Introduction

After some thirty years of research in the area of text-to-speech synthesis (Holmes, 1988) – that is, the automatic production of voice output from ordinary text input – we are at last beginning to be able to produce synthetic speech of high enough quality to ensure acceptability in a variety of products in the market place, including interactive applications (Morton, 1991). Although there are various techniques for generating high quality speech output in limited domain situations it remains the case that there is an ever growing requirement to produce the spoken word in environments where what is to be said is unpredictable. Text-to-speech synthesis is a method of producing spoken output from any text input. There is also an allied technique, called concept-to-speech synthesis (Young and Fallside, 1979), which is able to produce spoken output derived from inputs coming from devices which do not normally produce text. Apart from the nature of the input there is little difference between the two techniques that concerns us here.

Why have we had to wait so long?

We shall see in a moment that many of the problems encountered in designing a good text-to-speech synthesis system are by no means obvious. Because the greater part of the design of these systems was carried out in the early days of research, and has stood the test of time, researchers were unable to predict just how difficult the problem was going to turn out to be. This fact is true also of devices able to recognize the spoken word and convert it into text - automatic speech recognition systems (Holmes, 1988).

Text-to-speech synthesis is simply a simulation of the human process of reading text out aloud. Anyone can do this more or less adequately in their native language, and all that we have to do to engineer a synthesis system is determine what is involved in the human process and build a device to mimic this. Unfortunately determining what is involved in the human process has been much more difficult to accomplish than imagined. In fact, and perhaps ironically, building simulations has led us to realize just how little we knew of what is going on in the human process.

The early researchers in the 1960s relied heavily on what linguists and psychologists had to say about human speech production, since it is not possible to engineer anything successfully without having a good deal of reliable theory on which to build. It seemed self-evident that the specialists in language and speech would be in a position to supply some reliable information with which to underpin the simulation attempt - after all, linguistics had been around for some four thousand years. The linguists were happy to supply the information, though some of them were bemused that anyone would want to get a machine to read text. Remember, this was early on in the development of computers.

Ironically it was the fact that the information about the nature of language and how people speak was so readily available which ultimately held up the development of speech technology for such a long period of time. The information offered to those developing the simulations was wrong. Nobody at the time realized that the metatheoretical position adopted by linguists and psychologists was quite unsuited to the task in hand - the simulation of what a human being is doing. A metatheoretical position is simply the motive underlying what a theory builder does - and, since thirty years ago linguists could not have conceived of simulation as being a motive for their research into the nature of human language, such a goal could not have formed part of their motive. A theory has to be developed with a particular aim in mind, and different aims will produce different theories. The theory the speech engineers were getting was totally unsuited to the task in hand. It took around 25 years for this to be realized, and so it is only recently that simulation has become a motive or part of the metatheory for those engaged in finding out about how human beings read text aloud (Tatham, 1990).

This is not the place to go into the arcane details of linguistic and phonetic theory or to consider in detail, for example, the neuro-physiology of tongue movement or the aerodynamic and acoustic effects in the production of this or that piece of speech, but I will give one example which should illustrate the point clearly.

An obvious question to be answered when setting out to design any speech synthesis system is: What are the basic building blocks of speech? We need to know the size and nature of the units which are to be assembled to create the speech output. The answer seems self-evident. They are the sounds, which we can all make, which are strung together to form the words we speak: there are three of them in a word like *cat* and four of them in a word like *text* – or should that be three, or perhaps five? But if it is often not entirely clear how many sounds make up a word, how can we even begin to be clear as to their nature? The fact of the matter, or at least the theoretical position held since the late 1970s, is that each of these words, *cat* and *text*, consists of just one speech unit – what we used to call a single syllable. In this model a word like *synthesis* would consist of three such building blocks, *syn-the-sis*. There is a sufficient amount of evidence, accumulated since the start of speech technology research, to suggest that even at this most fundamental level in the problem, some of our earlier beliefs about the nature of speech were not quite right (Green *et al.*, 1990). Just changing the nature of the basic units to be conjoined to produce the output from a synthesis system makes a significant difference to the quality and naturalness of the output.

To create a convincing simulation of human speech production we need therefore to have at our disposal a clear theory of the natural process upon which to build a simulation model which will form the basis of our engineering solution.

Modelling reading text aloud

The input text

We begin a text-to-speech system with the assumption that we can input the actual text and assemble it in a form suitable for processing. This involves having the text in some electronic form, either directly from a keyboard or by some other means such as the signals involved in electronic mail transmission or videotex, or from some optical character recognition device scanning a printed page. From such inputs we are able to identify the individual words which make up the text (by means of the spaces occurring before and after them) and the individual phrases and sentences they form (by means of the conventions about the use of upper-case letters and punctuation marks). For the human being the input is by optical character recognition via the eyes and optical processing circuitry in the brain.

At this point human beings are involved in some language processing to determine the meaning of the text they are reading. We know this because we can observe that reading aloud often goes wrong if the reader does not understand what is being read, or begins a sentence assuming one meaning and then, say, turns a page to find out the assumption was wrong. The error is most often revealed in an incorrect (or unacceptable) rendering of the prosodics of a sentence. The prosodics include the rhythm of the sentence and its intonation, or general 'tune'.

Determining what the text means

Our simulation needs to determine what the text means. At the present time full scale understanding is impossible since we have not yet developed language processing technology sufficiently. But some extraction of the meaning is possible, however incomplete. Even the earliest text-to-speech systems would often attempt to determine the grammar of a sentence by identifying nouns and verbs, etc. This process is called **syntactic parsing** and is a small area of language processing where researchers have had success. The trick in a well engineered system is to determine just how much parsing is necessary and sufficient since full syntactic parsing still runs hundreds or thousands of times slower than real time. Syntactic parsing, though an essential part of understanding the meaning of text, is not enough; it will not be until we get some **semantic parsing** into our systems that text understanding will be a reality. With only one or two current text-to-speech systems even attempting the most rudimentary semantic parsing we can identify here an area where the simulation falls short of the ideal and in which errors will almost certainly be generated. The best systems include (as does the human being) techniques for minimizing the damage caused by such errors.

The dictionary

The parsing processes can be greatly assisted by including a dictionary in the system which can look up words in the text to determine, among other things, what grammatical category they belong to – that is, whether a particular word is a noun or a verb, etc. The dictionary can also contain information as to how a word might relate to other words; for example, you cannot associate the word *turquoise* with *misconception*. This latter kind of information assists the semantic parse. In early systems the dictionaries were kept to just a few tens of 'difficult' words because of the memory requirements for their storage and the problems associated with the time taken to look words up. Both these problems have been addressed in computer science and nowadays are no obstacle: dictionaries of over 100 000 words can now be contemplated. It will always be the case, though, that a word could appear in the text that is not in the dictionary, just as human beings will come across words when they are reading that they are not familiar with. In such a case the system just has to have a shot at reading, in much the same way as a human being does, perhaps even with the same kind of hesitation (or failure to keep the system running in real time).

Once we have a dictionary in the system we are free to include in it whatever information might assist in some of the processing to follow. Thus, just as in the dictionaries we use ourselves, we might find information about how to pronounce a word. English is full of words like *rough*, *cough*, and *bough* which it is impossible to know how to pronounce just by looking at the spelling. Early text-to-speech systems tried to include as many of these words as possible, but because of the memory limitations referred to above they had also to include rules for converting spelling to some kind of representation of the actual pronunciation. The vagaries of English spelling, however, are such that a complete set of rules is impossible since there are many hundreds of exceptions in the language. Clearly, the larger the dictionary the greater the chance of pronouncing any one word correctly.

Prosodics

If we know roughly what a sentence means and what its grammar is, and if we also know how the individual words are pronounced in isolation, we are in a position to look at the sentence as a whole and work out its prosodics. As mentioned earlier this means establishing a rhythm for speaking the text aloud and working out a suitable intonation.

The intonation will determine at what points the voice goes up and down in **pitch**. For example if we take the sentence *John went home* we can pronounce this with a pitch that starts at an average level and progressively falls, indicating that we are making a simple statement of fact. But if the sentence is pronounced with the voice progressively rising in pitch throughout then we know that we are asking the question *John went home?*–the changing pitch altering the meaning rather than any rearrangement of the words themselves. The situation is not all that simple – falling intonation for statements of fact and rising intonation for questions – because often questions have the falling pitch contour, as in *Why did John go home?*

With sentences of more than just three or four words the prosodics model is however extremely complicated. Human beings are constantly changing the rhythm and intonation even during the course of a sentence to emphasize the important words or sometimes just to add variety.

The simulation of the prosodic component in text-to-speech synthesis has proved to be the most difficult task in designing high-quality systems. This is partly a knock-on effect from errors in the earlier parsing of meaning (emphasis cannot be placed on important words in a sentence if the meaning is not known), and partly due to the prosodics model we have at the moment not being quite adequate. Significant progress has been made in this area in the last few years and one or two text-to-speech systems are able to reflect this progress in greatly improved speech output. Here, as elsewhere, the best results are obtained by those systems which take special care to minimize the effects of errors. When the system fails it must know it has done so, and take steps to produce an output which neither causes the listener to misunderstand nor cause him or her to smile (unless that is what is intended!).

Turning what to speak into how to speak it

At this point the text-to-speech system has worked out how it would like to pronounce the text it has 'read'. It knows how the individual words are to be pronounced, it knows what the sentences mean and it knows what kind of prosodics are needed to put all this across to the listener. This is analogous to a human being working out in his or her mind what to say before actually saying it – a process known to happen slightly before any speech sounds are made. In other words, the system has completed the planning stage of putting together what goes on behind the actual making of sounds. Now it has to turn those plans or intentions into actual soundwaves.

The devices for generating the actual waveforms are usually parametric in nature – that is, they create the waveforms from a finite number of component parts known to be the significant ones in speech. These parameters, usually a dozen or so in number, need continuous updating to produce the required soundwave. Our final task in a text-to-speech system is to assemble a file of numerical values for these parameters in real time. Such a file will take the form of a string of numbers for each of the parameters (let us assume there are twelve) to be issued every ten milliseconds to the device responsible for generating the acoustic signal; it then takes over and produces the speech independently.

So, the decisions taken by our text-to-speech system about how it would like to pronounce the text it is reading now have to be turned into a numerical control stream which will eventually result in an acoustic signal. The rather abstract intentions of the system about how it would pronounce the text now have to be converted into a numerical representation of what it will actually sound like.

The building blocks of speech

The task at this stage is to assemble the appropriate building blocks of speech into a sequence reflecting the system's intentions. The available building blocks are known to the system by being stored in an inventory of representations of all the speech units available in the particular language – in this case, British English. The quality of the resultant speech will depend on just what these units are and on how they are represented in the inventory.

Until very recently it was generally regarded that the building blocks of human speech were individual sounds, and that these were strung together to make up the words which in turn make up the sentences we wish to speak. Consequently most text-to-speech systems store in their building block inventories representations of these individual sounds which are usually taken to number between one and two hundred for any one language (Holmes, 1988). The word to focus on here is representations, since in such systems the sounds are not stored as waveforms but as a single set of values, one for each of the parameters of the device mentioned earlier which will eventually turn them into an acoustic signal. This is a very important point, because it means that the representation is of only a single, though typical, 10 ms worth of a sound that might eventually have a duration of perhaps 80 ms. In a sense this is an abstract representation which will form the basis of a process to derive a stream of parameter values for however many 10 ms frames are needed for the sound in question. The number of such frames normally required for a sound is indexed to its representation in the inventory.

One by one as required the parametric representations of the sounds are called from the inventory and multiplied up to give them their required durations. However there is a problem: when these sound representations are abutted there can often be a considerable discontinuity in the values for any one parameter at the boundaries between sounds (Holmes *et al.*, 1964). Such discontinuities do not occur in human speech because the sounds never occur in isolation; they blend continuously into each other as the tongue, lips, and so on, contribute in almost continuous movement to forming entire utterances – a process known as coarticulation. In a text-to-speech system this process is modelled by a smoothing procedure at the boundaries.

To clarify this let us consider three sounds, S_1, S_2 and S_3, in sequence, each with a durational index of five frames each of 10 ms. For a particular parameter we may get the following numerical sequence for sound S_1: 13 13 13 13 13, for S_2 the sequence: 25 25 25 25 25, and for S_3 the sequence: 18 18 18 18 18. When abutted we would get the stream:

... 13 13 13 13 13 25 25 25 25 25 18 18 18 18 18 ...

This sequence would be smoothed to something like:

... 13 13 13 **15 18 20 22** 25 **24 22 20 19** 18 18 18 ...

and so on. In human speech, although we do not get the discontinuities which the text-to-speech system would generate without some kind of smoothing, we do not in fact get such smooth curves as the usual algorithms produce. There is, in fact, a **systematic jitter** (if that is not a contradiction in terms) present in human speech, which has yet to be adequately modelled.

Because we do not yet understand this jitter (although we do know it to be systematic, since the introduction of random jitter does not produce the same effect perceptually) the solution in one or two recent text-to-speech systems has been to move away from the traditional type of unit and its representation I have just described to a syllable sized unit (as prescribed by recent phonetic theory) and toward a different representation in the inventory.

An improved representation of speech building blocks

Although we need some 200 units in the inventory if they are to be the individual sounds previously thought to make up speech, when we move to syllable sized units we need around 10 000 to exhaust all the possible syllables in a language. Once again, the lowering of storage costs and the development of efficient accessing algorithms permit us to contemplate such a system.

To make sure that the jitter present in human speech is incorporated into the system a different means of representation has been adopted. The basic idea of a parametric representation is still adhered to, but instead of an abstract representation of each syllable we move to what amounts to a copy of an actual syllable as produced by a human being. This is obtained by parametrically analysing the acoustic waveform produced by a person speaking and excising from it the individual syllables. It takes quite a lot of speech to get typical examples of every possible syllable. This parametric analysis of real speech is then stored as a complete entity for every syllable in the inventory.

The conjoining procedure simply extracts each required syllable unit in turn from the inventory, performing the minimum of smoothing at each syllable boundary. Thus the inherent jitter of human speech is preserved without having to be explicitly modelled.

Putting it all together

Once a sequence of speech units, properly conjoined, has been determined in the above stage there only remains the task of marrying this with the prosodic contour calculated earlier in the system. The basis of rhythm in speech is the sequencing of syllables, so a system which is syllabically based has already identified the necessary rhythmic units. A single-sound based system of the earlier type has, at this point, to identify the rhythmic syllables within the stream of sounds. Durations of syllables are carefully adjusted to match the rhythm requirements according to models of rhythm available in phonetics.

At the same time the intonation requirements generated in the prosodic component of the system are reinterpreted as a numerical string which is linked as a new parameter to the parameter stream already derived by conjoining inventory units. This process of reinterpretation of an abstract intonation representation is as yet not entirely satisfactory in any text-to-speech system generally available, and is too complex to discuss here. However new algorithms show promise by, once again, sensing errors and minimizing their effect.

The limits for natural-sounding synthetic speech

As far as the listener is concerned, natural-sounding synthetic speech is, by definition, indistinguishable from real speech. This does not mean that the synthetic speech is the same as real speech. The definition has been chosen to reflect the engineering concern of having to develop systems which have as their basis a theory of language and speech which is not yet up to the mark as a theory specifically developed for simulation purposes. The goal is to produce a simulation of the human output which is perceptually accurate by employing a system which is as good a simulation as we can manage of the human processes which derive that output.

The implication so far in this chapter is that we have recently moved suddenly very much closer to that goal - and indeed this is true for short stretches of synthetic speech that might be encountered in a voice operated information system working over the phone network. However for longer stretches of speech, or for continuous speech of a genuinely conversational nature - a dialogue with a computer - a machine-like quality is still able to be detected by the critical listener. Hence the question: What is it in human speech that is not adequately captured by the system so far described?

We know of two properties of human speech not yet incorporated in the text-to-speech systems I have described. They are

1. **variability** over stretches longer than a single unit (Tatham, 1989), and
2. a **pragmatic component** added to an otherwise neutral output (Morton, 1992).

Failure to incorporate either of these results in a less than natural-sounding output over long stretches of speech. We know that if we do incorporate them by synthesis of large stretches of speech derived entirely from parametric analysis of human speech – that is, not built up from the small units used in a text-to-speech system – the simulation gains a dramatic new dimension of naturalness. The question is how to model these two phenomena such that they can be incorporated into our text-to-speech system.

Variability in the articulation of speech units in stretches of speech is characteristic of all human speech. Current synthesis systems do not make provision for this kind of variability, with the consequence that repetitions are always rendered identically. A listener unconsciously detects this error and consequently feels the speech to be unnatural. The phenomenon is beginning to be modelled in Cognitive Phonetic Theory (Tatham, 1990); its explanation lies in the fact that a human speaker varies the precision of articulation depending on a predictive assessment of the listener's difficulty in understanding what is being said: if the speaker predicts the listener will encounter ambiguity or lack of clarity then the precision of articulation (and hence of the soundwave) will be increased and vice versa. In a synthesis system this would mean a continual adjustment to the 'accuracy' of the units retrieved from the inventory before conjoining them, dependent on the semantic, syntactic and phonological context of the units. This is not an easy task, and only one system (Lewis and Tatham, 1991) attempts such ongoing adjustments. It is able to do this by incorporating a model of human speech perception against which it tests every utterance it intends to make, adjusting the variability of the projected speech output to something more natural than would otherwise be the case.

Pragmatic effects are characteristic of every utterance in human speech. They are quite subtle effects overlaid on a normal neutral speaking 'tone' which convey to the listener such things as the mood of the speaker, his or her attitude to what is being said or his or her attitude toward the listener. In general such effects are most often encountered in changes to the prosodic element in human speech. Once again, there is only one text-to-speech synthesis system (Lewis and Tatham, 1991) which attempts to generate these effects – with the result that the listener feels he or she can detect the speaker's feelings.

Characterizing the prosodic effects which communicate a speaker's feelings has proved a difficult task, and the best results have been obtained from training a neural network to learn the effects for itself by presenting it with many examples of human speech. The neural network is then used to modify the otherwise pragmatically neutral output of the text-to-speech system. Both these phenomena will undoubtedly be incorporated in text-to-speech synthesis systems in the future.

The one system which has demonstrated that it is viable to include them has also shown the dramatic increase in naturalness which they produce (Lewis and Tatham, 1991). This is the SPRUCE text-to-speech synthesis system being currently developed in a joint project between the Universities of Essex and Bristol (Fig. 3.1). The system, though still in its early stages of development, has been able to demonstrate that by incorporating the recent advances in basic theory (such as the use of parametrically analysed human syllables) and paying attention to such factors as variability and pragmatic effects a naturalness of speech output quality, hitherto not obtained is quite possible.

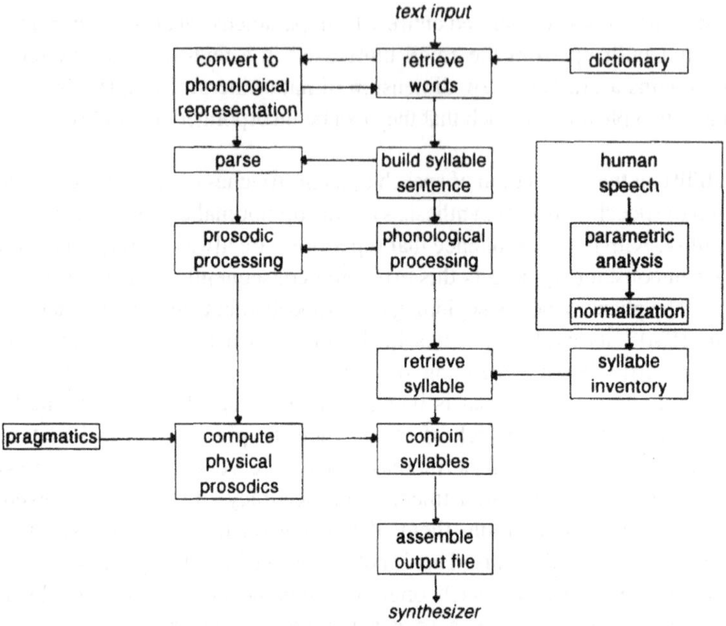

Figure 3.1 SPRUCE text-to-speech synthesis

Future systems

What can be realistically expected of future text-to-speech synthesis systems, and when can we expect to find them ready for the market place or ready to be incorporated into other products needing voice output? Wary of the fact that researchers have been predicting the imminence of high-quality synthetic speech for around thirty years now, I find it dangerous to make very firm predictions, but it is worth examining why it may just be possible to be a little clearer now.

To begin with, we appreciate now that the developers of early systems, through no fault of their own, had to rely on theories of speech production which had not been designed for the task in hand. That situation is now changing, and in the last few years we have seen the beginnings of a more appropriate theory designed for supporting the modelling of simulations of speech production. Given this impetus, together with the demonstration of the considerable advantages such changes in the theory bring, we should see much better systems delivering very natural speech in the next five years or so.

There is one other factor which affects the timing of the availability of natural-sounding text-to-speech synthesis. Over the past ten or fifteen years there was a widespread view that text-to-speech synthesis had been largely solved because at least the speech from the devices then available was intelligible. Many researchers and the funding agencies they depended on consequently dropped the work in favour of less well developed areas of speech technology such as automatic speech recognition which was lagging well behind synthesis. What those who judged that the task of building synthesizers was all but finished failed to get right was the lay user's reaction to what was available: it turned out that users were not impressed at all by text-to-speech synthesis – they want intelligibility of course but they also want naturalness, and that was not available. However, now that we have had considerable advances in the field of automatic speech recognition, attention is turning once again to speech synthesis, and perhaps the investment that is still necessary in this field will be forthcoming. Certainly we can now show, with SPRUCE for example, that such investment is worthwhile and that it will produce the desired improvements in text-to-speech synthesis very quickly.

References

Green, P. D., Brown, G. J., Cooke, M. P., Crawford, M. D. and Simons, A. J. H., 1990, Bridging the gap between signals and symbols in speech recognition, in *Advances in Speech, Hearing and Language Processing*, Vol. 1 W. A. Ainsworth (Ed.), London: JAI Press.

Holmes, J. N., 1988, *Speech Synthesis and Recognition*, Wokingham: Van Nostrand Reinhold.

Holmes, J. N., Mattingly, I. G. and Shearme, J. N., 1964, Speech synthesis by rule, *Language and Speech*, **7**(3).

Lewis, E. and Tatham, M. A. A., 1991, SPRUCE – a new text-to-speech synthesis system, *Proceedings of Eurospeech '91*, Genoa: ESCA.

Morton, K., 1991, Expectations for academic and industrial co-operation in linguistics processing, *Proceedings of Linguistics Engineering '91*, Nanterre: Editions Colloques et Conseil.

Morton, K., 1992, Pragmatic phonetics, in *Advances in Speech, Hearing and Language Processing*, Vol. 2 W. A. Ainsworth (Ed.), London: JAI Press.

Tatham, M. A. A., 1989, Intelligent speech synthesis as part of an integrated speech synthesis/automatic speech recognition system, in *The Structure of Multimodal Dialogue*, M. M. Taylor, F. Néel and D. G. Bouwhuis (Eds), Amsterdam: North Holland.

Tatham, M. A. A., 1990, Cognitive Phonetics, in *Advances in Speech, Hearing and language Processing*, Vol. 1 W. A. Ainsworth (Ed.), London: JAI Press.

Tatham, M. A. A., 1990, Preliminaries to a new text-to-speech synthesis system, *Proceedings of the Institute of Acoustics*, **8**, London: Institute of Acoustics.

Young, S. J. and Fallside F., 1979, Speech synthesis from concept: a method for speech output from information systems, *Journal of the Acoustical Society of America*, **66**.

4

Interactive speech in computer-aided learning

Eric Lewis

Abstract

Computers first began to be used as teaching aids in the 1960s. At that time the facilities were rather crude by today's standards but as the technology has advanced computer based training/computer-aided learning (CBT/CAL) systems have taken advantage of the developments so that now an individual user can be provided with a powerful desktop computer and a reasonably 'friendly' human–machine interface. However, this interface has so far failed to take advantage of the most fundamental means of communication, namely speech, which is contrary to most other modes of teaching in which speech is accepted as the main medium of communication between teacher and pupil. This chapter examines the different ways in which interactive speech can be provided and argues that the technology is now available for speech to play a more significant role in CAL/CBT systems than has previously been the case.

Introduction

Teaching and learning are based on communication, and communication between humans relies on the five senses - touch, taste, smell, sight and sound. Of these it is arguable that sight and sound are the most important and that sound is the most important of all since speech is the most fundamental means of communication. Education at all levels relies heavily on the spoken word and it is difficult to think of any teaching practices (other than those involving computers) which do not involve speech. In some instances, speech is almost the only mode of communication possible, for example pre-school children are usually unable to read and blind people must rely on speech until they learn Braille. Even when circumstances dictate that speech communication is difficult, for example in distance learning, every effort is made to provide the student with occasional face-to-face tutorials or audio back-up. As students move up the academic scale they are certainly expected to participate in more self-study but at no stage are they expected to learn without any face-to-face tuition and that means a spoken dialogue with a tutor or lecturer. The inescapable conclusion, therefore, is that speech plays a very important part in the teaching and learning process.

Computer-aided learning (CAL) is a concept that started in the US in the 1960s and initially was identified with programmed learning. CAL has subsequently taken on a much broader meaning and covers the use of computers as a learning resource, whether it be in schools, universities, industry or commerce. It is not the intention to debate the merits, or otherwise, of CAL here but rather to argue for an improvement in the human-computer interface for CAL systems. At first the facilities for CAL were very basic consisting mainly of minicomputer systems linked with teletype terminals. The facilities have now advanced to the stage where users can have their own personal computer consisting of a powerful microprocessor, medium to high resolution colour monitor, keyboard and mouse together with a WIMP interface which makes the computer considerably more 'user-friendly'. Over the last few years high quality graphics has been possible which enables CAL software to provide material that could previously only have been provided on film or video. Currently the CAL environment is being enhanced with multimedia systems which combine computers with computer controlled videodiscs and CD-ROMS, enabling high quality images, with accompanying sound or speech, to be displayed on the screen. In addition authoring systems are being developed which will enable CAL designers to construct courseware which incorporates the full range of multimedia facilities.

Speech with CAL

What are the advantages of using speech with CAL? They are surely the same as those for using speech in existing traditional teaching methods. Speech provides a further channel of communication which can be used to advantage. Consider the number of occasions when one wishes to point to an overhead projector transparency or the blackboard and describe some particular feature. The pointing mechanism is easily reproduced with a computer but currently the description is usually provided in text on some other part of the display screen or in accompanying documentation. This distracts the user from the feature being described. There are many examples where this situation occurs.

Distance learning has been with us for a long time and has been very successful with the Open University. Extensive use of audio cassette tapes already takes place with the OU which clearly indicates that the OU considers communication via speech to be important. The OU is also investigating the provision of a purpose built computer called the Thought Box. Its architecture is described in an article by Alexander and Lincoln (1989) in which they state that

> the combination of speech and visuals is likely to be the most powerful computer based learning medium of short to medium term.

One should also not forget how computers have been able to improve the environment for the disabled. Learning opportunities for the blind and other disabled people could be substantially improved by the use of interactive speech.

There are, of course, disadvantages in using speech. One which is frequently mentioned is that speech does not provide the user with a hard copy of the dialogue. If the meaning of the speech is not grasped then it has to be repeated. This situation occurs frequently in speech dialogue without noticeably causing any problems so it is not really a significant handicap. If it is really important that the speech exists in hard copy for subsequent perusal then it can always be printed on the screen as well as spoken. The fact that the speech is transient may even, perhaps, result in the user applying more concentration to the dialogue with a consequent increase in understanding.

Speech is by its very nature obtrusive and while this has benefits in as much as the user's attention is demanded it means that the computer systems have to be equipped with headphones and directional microphones or else placed in sound-insulated enclosures. There are, therefore, environmental constraints which may be difficult to implement.

If speech had been available with computers from the beginning then there can be no doubt that speech would have been provided as an integral part of CAL software. The justification for this statement lies in the fact that with all the different media that are used in education, none of the available channels of communication are neglected. Film and video, for example, rarely if ever display material that uses only images, though occasionally the accompanying audio may be just sound with no speech. Now that multimedia workstations are becoming available it will surely be taken for granted that their speech production facilities will be used just as frequently as their image processing capabilities.

Previous use of speech in CAL

In view of the advantages of speech just described and the importance of speech as a communication channel it would seem that efforts must have been made to introduce speech into CAL. This is in fact the case.

In the early 1980s the only viable means of providing speech with CAL was by means of audio tape recorders. It was possible to control these recorders using the computer but the process was rather cumbersome and the necessarily linear searching of the tape made it difficult for cue and review operations (Miller, 1983; Miller *et al.*, 1983). Specialty tape recorders were developed (Tandberg, 1983; Curtis, 1984) which provided high speed searching but even so they have not led to the extensive use of audio tape recorders for CAL packages. Towards the mid-1980s researchers started experimenting with synthetic speech (Gray, 1984; Gull, 1985) and concluded, 'that synthetic sound (particularly speech) can significantly enhance the quality of CAL'. Lewis and Buxton (1991) have experimented with using synthetic speech for a spreadsheet tutorial and their results seem to indicate that a speech based tutorial is more effective than a text based tutorial, though ideally the subjects would have preferred a combination of text and speech communication. It is difficult to find many attempts to introduce synthetic speech in CAL and this is most probably due to the fact that the improvement in quality of synthetic speech over the last five to seven years has not been sufficiently good. Leedham (1991) comments that, 'speech synthesis can be achieved by a number of techniques but the quality of currently available speech synthesisers is generally poor and none sounds natural'. Recognizing this fact, Bridgman (1990) used a Votan voice card capable of storing speech in a compressed digital form to develop CBT software utilizing interactive voice. His objective was to investigate the practical application of audio in CBT and he tentatively concluded that interactive voice had been instrumental in producing a user-friendly CBT program.

The use of audio tapes is still considered to be a viable form of teaching by the OU (Jones *et al.*, 1992). A recent use of this teaching medium has involved providing audiotape tutorials for novice MS-DOS users who have their own home computer, and their experience suggests that guided audio introduction is necessary for all OU home-based computing users. How much more convenient it would be, however, if the speech could be integrated with the tutorial package and not involve the simultaneous manipulation of tape recorder and computer.

For some time Interactive Video has held out promise of considerably enriching the CAL environment by providing high quality images as well as sound. However, it has not progressed as rapidly as one might have expected and a recent report (Norris *et al.*, 1990) on the Interactive Video in Schools (IVIS) project documented several impediments to its take up in secondary schools even though they concluded that, 'there is little doubt that teachers and pupils judge Interactive Video to be an important addition to the learning resources of a school'. The production of videodiscs is a time consuming and expensive process and is unlikely to be economically viable in education unless there is a large take up of the finished product. This would require a significant investment in equipment by schools and is unlikely to take place in current circumstances.

The rapid development of faster processors coupled with the improved hard disc technology has enabled facilities for the digital recording and subsequent playback of speech to be provided with standard computers. The Macintosh has led the field in this respect and using its Hypercard system it is relatively easy to develop CAL software incorporating speech. Davidson *et al.* (1991) have used such a platform to produce talking books to assist primary school children in learning to read. Benest (1992) has used a SUN workstation in a similar manner to produce talking books for use with undergraduates in a university environment. The advantage of digital recording is, of course, that the speech can be stored on hard disc along with any other data and can be rapidly searched for playback. There is one major disadvantage, however, to storing speech on hard disc and that is the amount of space it occupies. For reasonable quality speech about 600K bytes are required for just a minute of speech and half an hour would require nearly 20M byte. This figure can be reduced by using compression techniques but even so it will be necessary for speech to be stored on CD-ROM if significant amounts are to be used. Multimedia workstations possess just such a facility.

The applications above refer only to speech output and not speech input. While some published results for studies concerning the use of speech driven interfaces claim that speech input offers no significant advantage over other standard input devices (Gould *et al.*, 1983; Brandetti *et al.*, 1988), others claim that speech input is more efficient than typed input and provides an additional response channel extending the user's efficiency (Martin, 1989). The conclusion would seem to be that the effectiveness of speech input depends upon the task specified. In a CBT environment Bridgman's (1990) attempts to use speech input for development of the Microtext Talking Tutor were disappointing and he was unable to develop any useful design criteria. However, SRI International appear to have had some success in developing a Voice Interactive Language Instruction System that uses audio input and output for foreign language teaching (Bernstein and Rtischev, 1991). In general speech recognition technology is not yet at the stage where it can cope with speaker independent continuous speech and therefore its use in a CAL environment is still severely limited.

Future provision of interactive speech

Given that speech is required as part of the human machine interface the next question to be answered is how can its use best be implemented. The easiest way to provide speech output is almost certainly by digitally recording real speech. The provision of recording and playback facilities is rapidly becoming standard on all types of workstations. The newest Macintosh computers already come with sound input/output hardware and software while relatively inexpensive add-on cards are available to upgrade PCs in the same way. In fact audio hardware is now part of any multimedia system and any workstation is capable of being enhanced to become a multimedia system. There are a number of workstation vendors who are strongly advertizing their computing platforms as being ideal for providing multimedia capabilities. No-one would consider buying a computer today without decent graphics facilities and it is no exaggeration to say that within a few years no-one will consider buying a computer without decent multimedia facilities. It would appear, therefore, that the hardware is now in place for CAL software to be developed which fully integrates graphics, video, text, sound and speech. CAL will therefore be able to deliver learning material whose appearance is as sophisticated as film or video but with the added advantage of allowing the user to interact with the medium.

Looking more to the future the advantages of using synthetic speech will surely mean this form of speech output will replace much digitally recorded speech. Currently synthetic speech does not have sufficiently good intonation but this drawback will undoubtedly soon be corrected (Lewis and Tatham, 1991). Synthetic speech produced from text will have the advantage of being extremely economical on storage, one minute of speech occupying about 1K bytes. Furthermore, the CAL designer will easily be able to edit the speech, change voice and change emphasis providing a flexibility which will give unrestricted text-to-speech synthesis a major advantage for computer controlled speech output.

For input the interaction between user and computer is mainly provided by keyboard or mouse. Limited speech input in specific applications is proving to be successful but ideally one is looking for completely speaker independent continuous speech recognition and that has still to be achieved in the marketplace. The development of speech input is not sufficiently far advanced for speech recognition systems to be incorporated as part of the standard multimedia platform.

Conclusions

The following quotations are taken from Barker (1986a, 1986b):

> Many computer-based training and learning situations require audio support for their effective implementation.
> Good quality speech can now be generated relatively inexpensively by computer.
> There are many avenues along which future developments in HCI/MMI are likely to proceed. The more obvious include greater utilisation of speech processing and image analysis...
> In many CAL situations the use of sound (through interactive audio) is a necessity.

It would seem that Barker's enthusiasm for the use of speech in CAL has not been echoed by many of his colleagues since the inclusion of speech in CAL software is still relatively rare. However, the main stumbling block has been the lack of easy-to-use facilities for producing speech output and recognizing speech input. As far as speech output is concerned that block has now been removed with the arrival of multimedia workstations and it is inevitable that producers of all software, not just those involved in CAL, will seek to use the speech facilities whenever speech would form a natural part of the interface. This view is confirmed by Maddix (1990) who states that

> The basic motivation for using multimedia interfaces stems from the improved effectiveness and efficiency of communication that can be achieved by using several channels of communication simultaneously in parallel with each other.

At present, and for the next few years, speech output in CAL will be obtained by digitally recording and compressing human speech. However, once text-to-speech synthesis can reproduce the naturalness of human speech, it will play a much more significant part due to its inherent flexibility compared with recorded speech.

The timescale for the inclusion of speech input in CAL systems is longer but the arrival of speaker independent continuous speech recognition on the scene will mean that speech will then be able to play its full part in the human–computer interface.

The emphasis now should be to develop the authoring systems which will allow designers of CAL software to incorporate speech easily into their programs. It is important also that such designers incorporate speech into the design of their software from the beginning and do not try just to add speech to their existing interfaces. It is worth remembering one of Leedham's (1991) conclusions that while speech is a more natural form of communication than other commonly used devices it does not offer, 'a panacea to human interface design'.

References

Alexander, G. and Lincoln, G., 1989, *Mindweave: Communication, Computers and Distance Education*, Pergamon Press, Oxford, England.

Barker, P., 1986a, The many faces of human-machine interaction, *Br. J. Ed. Tech.*, **17**(1), 74–80.

Barker, P., 1986b, A practical introduction to authoring in Computer Assisted Instruction, Part 6; interactive audio, *Br. J. Ed. Tech.*, **17**(2), 110–28.

Benest, I., 1992, Towards computer based talking texts, *Conference on the Teaching of Computer Science*, Kent University, April.

Bernstein, J. and Rtischev, D., 1991, A voice interactive language instruction system, *Proc. Eurospeech '91. 2nd Eur. Conf. Speech Comm. & Tech.*, 981–4, Genova, Italy.

Brandetti, M., D'Orta, P., Ferretti, M. and Scarci, S., 1988, Experiments on the usage of a voice activated text editor, *Proc. of Speech '88, 7th FASE Symposium, Edinburgh*, 1305–10.

Bridgman, T. F., 1990, Applied interactive voice in CBT: A pilot project, *Computers & Education*, **15**(1-3), 173–81.

Curtis, M., 1984, Tandberg TCCR-530 computer programmable cassette recorder, *Soft Options*, 14–16, July.

Davidson, J., Coles, D., Noyes, P. and Terrell, C., 1991, Using computer delivered natural speech to assist in the learning of reading, *Br. J. Ed. Tech.*, **22**(2), 110–18.

Gould, J.W., Conti, J. and Hovanyecz, T., 1983, Composing letters with a simulated listening typewriter, *Comms. ACM.*, **26**, 295–308.

Gray, T., 1984, Talking computers in the classroom, *Electronic Speech Synthesis: techniques, technology and applications*, 243–59, G. Bristow (Ed.), Granada, London.

Gull, R. L., 1985, Voice synthesis: an update and perspective, *Proc. IFIP WCCE 85 Fourth World Conference on Computers in Education*, 525, K. Duncan and D. Harris (Eds), Norfolk, Virginia, July.

Jones, A., Kirkup, G., Kirkwood, A. and Mason, R., 1992, Providing computing for distance learners: a strategy for home use, *Computers & Education*, **18**(1-3), 183–93.

Leedham, G., 1991, Speech and Handwriting, in *Engineering the Human-Computer Interface*, A. Downton (Ed.), McGraw-Hill, UK.

Lewis, E., Buxton, D., Giles, A. and Ooi, J. C. P. and Buxton, D., 1991, The use of speech synthesis in Computer Aided Learning, Report TR-91-35, Department of Computer Science, University of Bristol, Bristol, BS8 1TR.

Lewis, E. and Tatham, M. A. A. T., 1991, A new text-to-speech synthesis system, Proc. Eurospeech '91, 2nd Eur. Conf. Speech Comm. & Tech., 1235–8, Genova, Italy.

Maddix, F., 1990, *Human-Computer Interaction. Theory and Practice*, Ellis-Horwood, NY.

Martin, G. L., 1989, Utility of speech input in user-computer interfaces, *Int. J. Man Machine Studies*, **30**(4), 355–75.

Miller, K., 1983, A development aid for audio-enhanced Computer Assisted Learning using the BBC microcomputer, Dept. of Physical Sciences, Wolverhampton, WV1 1LY.

Miller, L., Norton, K., Reeve, R. C. and Servant, D. M., 1983, Audio-enhanced Computer Assisted Learning and computer controlled audio-instruction, *Computers & Education*, **7**(1), 33–54.

Norris, N., Davies, R. and Beattie, C., 1990, Evaluating new technology: the case of the Interactive Video in Schools (IVIS) programme, *Br. J. Ed. Tech.*, **21**(2), 84–94.

Tandberg, 1983, Tandberg TCCR-530 computer controlled cassette recorder–characteristics and applications, Tandberg Ltd., Revie Road Industrial Estate, Elland Road, Leeds, LS11 8JG.

5

Speech-based alarm displays

Neville Stanton

Introduction

Speech synthesis is receiving increasing attention as a potentially valuable medium for a variety of applications of information display. This chapter addresses the use of speech as a means of displaying alarm information in 'process control' type tasks. Typically the operator of a control desk will monitor the state of a process from a room that is sited remotely from the plant. Therefore the operator's main source of information about the plant status is via process page displays presented on visual display units (VDUs), which may contain up to 800 pages of information. Further, the process plant is controlled by a team of operators and a supervisor who are in contact, either face-to-face or via telephone, with plant engineers.

A speech-based display medium might have potential benefits to offer control room operation. Benefits that are often associated with speech displays include: breaking through attention, eyes free/hands free operation, omnidirectionality, no learning required, reduction in visual clutter display. It has also been suggested that auditory channel is particularly well suited to the transmission of warnings (Stokes *et al.*, 1990). Therefore given the demands placed upon the operator in the control room, communication of alarm information using the auditory channel might present a way in which better use may be made of their limited attentional resources.

The use of auditory displays in control rooms is not a new idea, in fact most control rooms employ non-speech auditory displays in conjunction with visual displays for conveying alarm information. However, non-speech warnings are clearly very limited in terms of the amount of information that can be transmitted and the number of different types of signal which a human can discriminate. Patterson and Milroy (1980) suggest that it is relatively easy to learn up to seven different tone-based auditory warnings, but over this number becomes much more difficult. Thus speech might be a more flexible and informative medium than tone warnings, as this can be used to alert the operators to the problem, inform them of the problem's nature and cue the required response.

However, Baber (1991a, 1991b) warns that although synthesized speech appears to be an atractive display medium for the presentation of warnings in control rooms, one needs to consider the appropriateness of speech for the tasks performed in the application domain before it may be recommended. Baber (1991a) also presents some design considerations regarding warning systems, i.e. he suggests that the warning should sound distinct from human speech and that the message should be worded as a short phrase containing a minimum of five syllables. These recommendations are intended to increase intelligiblity and inform the operator that the message is from the machine, not another human operator.

In an experimental study reported by Baber *et al.* (1992) it was suggested that there might be a place for speech-based alarm displays in control room activities. In particular they proposed that a major benefit of speech in information processing terms was that it could be used to relieve the visual channel. This was illustrated by introducing Wickens' (1984) model of information processing which suggests that visual-spatial information and auditory-verbal information draw on separate 'pools' of attentional resources. Therefore transferring the alarm system's information from the visual display to an auditory channel might provide the human operator with greater capacity to deal with the incoming information and deal with it appropriately. This could be seen as a possible solution in attempting to spread the mental workload demands of the task. Thus in information processing terms, speech-based alarm systems might reduce the attentional demands on the visual channel. However, Baber *et al.* (1992) caution interpreting Wickens' (1984) multiple resource theory without first subjecting it to the rigour of the experimental laboratory.

In a discussion of intrinsic qualities of various alarm media, Stanton *et al.* (1992) illustrated that speech-based alarm and scrolling text alarms have much in common. Their presentation is temporal in nature, they contain complex information, they tend not to be grouped and the information takes the form of a message (containing information such as: plant area, plant item and nature of fault). There are also three major differences: the channel, means of access and the duration of the message. The 'channel' refers to the modality of communication (i.e. auditory or visual), the 'means of access' refers to the way in which the information is called up by the operator, and the 'duration' refers to the time for which the information is present. If the two media are to be separate then the channel and access are likely to remain different. Duration is another matter. Typically scrolling text messages are displaced by new incoming messages, but they can be called up again, and therefore may be referred to as semi-constant. Speech messages on the other hand, are transitory in nature. This means that as soon as a message is finished it is gone and cannot be retrieved. One means of making the messages more permanent could be for them to continue until they are dealt with, but that might lead to 'auditory clutter' and the temporal information may be lost.

In the previous study comparing human and synthesized speech-based alarms, Baber *et al.* (1992) proposed that speech-based alarm display needs to be considered in terms of the tasks that operators are required to perform. For this purpose a model of alarm handling was taken into the experimental laboratory, so that the medium could be investigated in a more rigorous manner. It was proposed that synthesized speech was more appropriate than 'human-like' speech, to maintain a distinct difference between information communicated from the displays and information communicated between operators. It was also suggested that speech might only be suited to a limited range of display information.

Therefore the research to be reported next builds upon the study by Baber *et al.* (1992) to investigate the operational performance differences in an experimental 'process control' task by comparing speech and traditional scrolling text alarm displays. It was proposed that speech-based displays would be superior for attracting attention and single fault management (i.e. one alarm connected to one event), whereas scrolling text displays would be superior for more complex multi-alarm problems (i.e. many alarms connected to one event). The combination of speech and text should therefore lead to superior performance overall.

Method

Participants

Thirty undergraduate students aged between twenty and twenty-four years took part in this study. The participants were allocated to one of three experimental conditions. Each condition contained five females and five males.

Design

The experiment consisted of three phases: training, data gathering and a surprise recall test. All participants went through the same phases, only the data gathering was different depending upon the experimental condition: speech alarms, text alarms or speech and text alarms.

Equipment

Training was presented on a Sharp 14-inch colour television running from a Ferguson Videostar VCR. The experiment was run on an Archimedes 310 microcomputer presented via a Taxan 770 colour monitor utilizing mouse and keyboard for input. Synthesized speech was generated through phonemes created on a synthesis-by-rule software program called 'Speech!' Synthesis-by-rule was chosen over pre-recorded messages for two main reasons. First, the alarm message should be distinct from human speech. Second, in a real human supervisory control task there could be up to 20,000 alarms (such as in a nuclear power station). It would be a daunting task to pre-record all 20,000 messages, whereas synthesis-by-rule offers a practical solution. The surprise recall task used a pencil and paper so that participants could record the alarms they recalled.

Procedure

The procedure for the experiment was as follows:

1. Participants were presented with a ten minute training programme on the television and VCR.

2. Participants practised the task (described below) on the simulator until they reached criterion performance of 80 per cent 'output'.
3. Participants were then read a set of instructions by the experimenter to explain that the plant would be masked so that they would have to use the 'inspect' keys to call up the status of a plant variable.
4. Participants continued in the experimental phase in which they encountered an unpractised emergency.
5. Participants were de-briefed about the nature of the study for five minutes.
6. Participants were presented with a surprise recall task, in which they had to recall as many alarms as possible.

Task

Briefly, the experimental task required participants to conduct a planned activity in a simulated 'process' plant. They had control over valves (open or close) and boiler heating (off, low, medium, high or very high). They could also inspect the status of the plant elements, e.g. tank levels, valve positions and boiler temperature. Using this information they were required to heat up a liquid until it was within predefined thresholds, then they had to condense it whilst it was inside these thresholds. Several elements of the process could go wrong. For example: the source liquid could run out, the supply pipe could crack, the temperature of the boiler could be too hot or too cold, or the coolant tank could run out. Each of these problems had an associated alarm. The subject's goal was to 'process' as much liquid as possible to achieve the maximum 'output'. In addition, the participant was requested to attend to a spatial secondary task when the workload on the primary 'process' task permitted.

Measurement

Every input the participant made was logged automatically by the computer. In addition the alarms generated and the 'process output' were also logged. This generated much data for each participant. Overall measurements were taken of output performance. In addition, participant response times to accept alarms, diagnose faults and recover the process were logged. These same measurements were used for collection of fine grain information in the unpractised emergency. Other data collected included: inappropriate activities, secondary task and the surprise recall task.

Analysis

Output performance and response times were analysed using analysis of variance followed by the Scheffé F test for *post hoc* analyses where appropriate. The data on inappropriate activities and recall performance were analysed using Kruskal-Wallis followed by Mann Whitney for *post hoc* analyses where appropriate.

Results

The main results show that there were some statisical differences between the experimental conditions. These are indicated below.

Output performance: $F_{2,27}=3.744$, $p<0.05$.
Accept alarms response time: $F_{2,25}=7.418$, $p<0.01$.
Pipe break accept response time: $F_{2,24}=21.986$, $p<0.0001$.
Pipe break recovery response time: $F_{2,25}=4.916$, $p<0.025$.
Inappropriate actions: H corrected for ties=*8.661*, $p<0.025$.
Secondary task: H corrected for ties=3.435, p=not significant.
Recall task: H corrected for ties=11.571, $p<0.01$.

The results of the *post hoc* analysis are indicated in Table 5.1 below.

The results generally suggest that performance was better in the text and speech and text (S&T) conditions when compared with the speech only condition. No statistical differences were found between the text and speech and text conditions.

Figures 5.1 to 5.5 illustrate where statistical differences were found. The figures show output performance (Fig. 5.1), common alarm activities (Fig. 5.2), pipe break activities (Fig. 5.3), inappropriate actions (Fig. 5.4) and performance on the recall task (Fig. 5.5).

As Fig. 5.1 shows, output performance was significantly worse in the speech-based alarms condition than in the other two experimental conditions (Table 5.1).

As Fig. 5.2 shows, time to accept alarms was significantly slower in the speech-based alarms condition than in the other two experimental conditions (Table 5.1). Although not statistically significant, participants in general in the text condition appear faster than the participants in the other two conditions for the investigative and recovery tasks.

As Fig. 5.3 shows, acceptance time was significantly slower in the speech-based alarms condition than in the other two experimental conditions (Table 5.1). Although not statistically significant, the participants in the speech only condition appear slower

Table 5.1 Post hoc *analyses of experimental conditions*

Dependent variable	Text vs. S&T	Speech vs. TEXT	Speech vs. S&T
Output performance	NS	*	NS
Accept alarms	NS	*	*
Diagnose fault	NS	NS	NS
Recover process	NS	NS	NS
Pipe break accept	NS	*	*
Pipe break diagnose	NS	NS	NS
Pipe break recover	NS	NS	*
Inappropriate actions	NS	**	*
Secondary task	NS	NS	NS
Recall task	NS	***	**

(where *=0.05, **=0.01, ***=0.001)

than participants in the other two conditions in the investigative task. Participants in the speech only condition were significantly slower than participants in the speech and text condition in the recovery task.

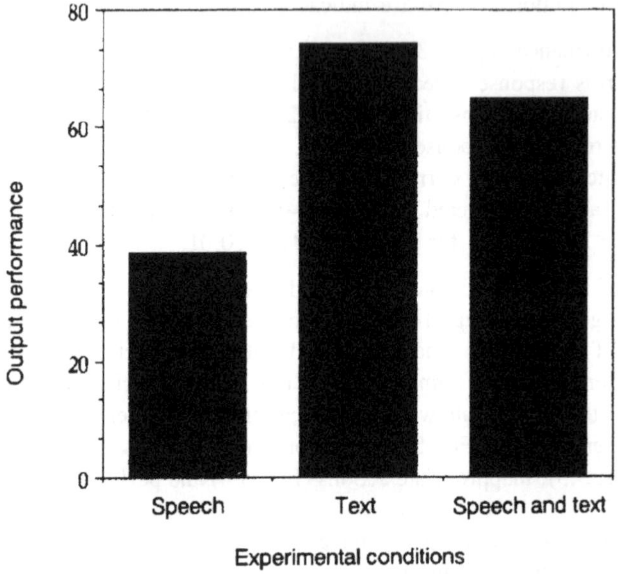

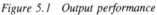

Figure 5.1 Output performance

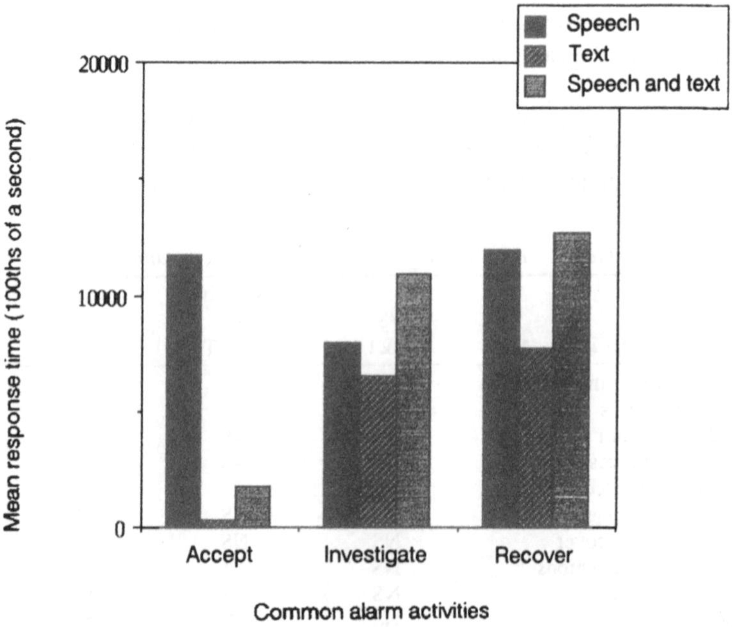

Figure 5.2 Common alarm response times

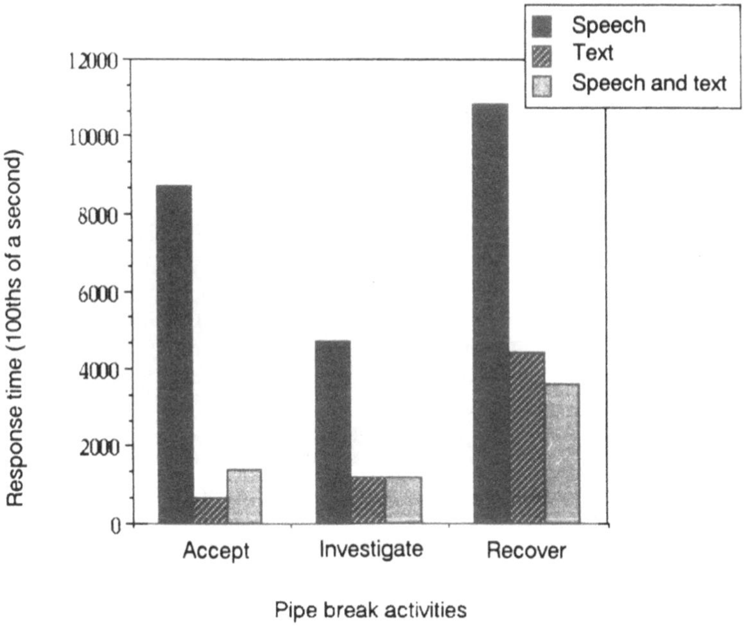

Figure 5.3 Pipe break response times

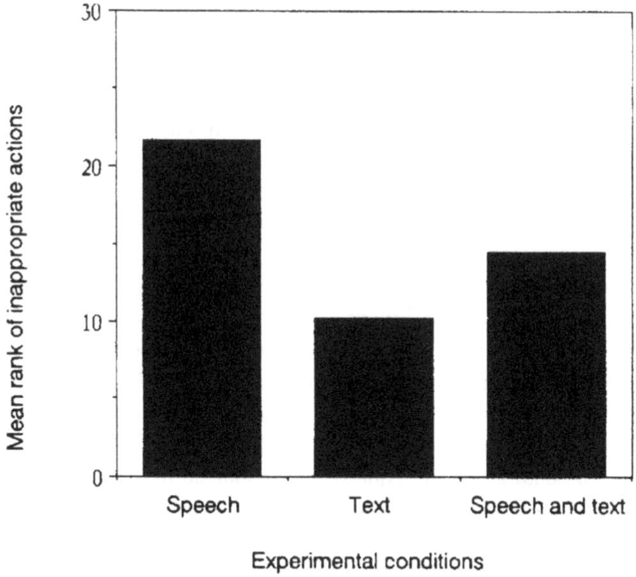

Figure 5.4 Inappropriate actions

As Fig. 5.4 shows, participants in the speech condition carried out significantly more inappropriate actions than participants in the other two conditions (Table 5.1).

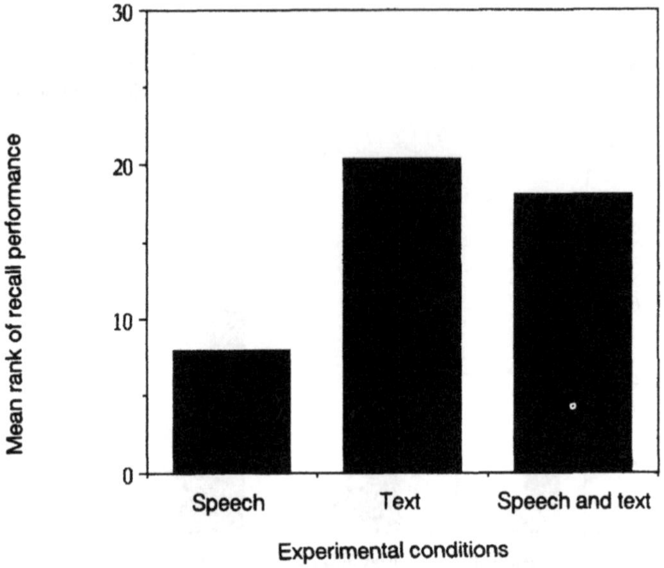

Figure 5.5 Recall performance

As Fig. 5.5 shows, participants in the speech-based alarms condition recalled significantly fewer alarms than participants in the other two conditions (Table 5.1).

Discussion

Interestingly, participants' performance in the speech only condition was generally significantly worse than that of participants in the other two conditions. This effect appeared to reduce substantially when speech was combined with a scrolling text alarm display. Thus one must consider why speech based alarm displays appear to be detrimental to performance in a 'process control' task.

It has been suggested that the presentation of verbal information is inappropriate for fault diagnosis (Robinson and Eberts, 1987). The suggestion is that the operator isolates the plant component in terms of its spatial reference rather than its verbal reference. Alarm information in the experimental task described within this chapter was verbal in all three conditions, but the duration of the information was much shorter in the speech-based alarm condition. This is a more likely explanation of the findings, as all conditions were equally disadvantaged in 'spatial reference' terms, i.e. in all of the conditions alarm information was presented verbally, as speech, scrolling text or both. Baber (1991a) also points out that poor quality synthesized speech and the resulting high memory demands this makes on operators could degrade performance. The latter of these probably had the greatest effect in this investigation. A discussion will expand on this later.

The results from the surprise recall task show quite clearly that participants in the speech-only condition were unable to recall as many alarm messages as participants in the other two conditions. This supports the findings of the first studies that report recall performance for a synthesized speech condition was very poor. This suggested that synthesized speech is processed at a surface level, rather than at a semantic level, which could account for the poor recall performance. Thus one could surmise that synthesized speech is inappropriate for tasks that have a memory component.

Speech also has a 'durational' component which may lock people out of the interaction until the message is complete. This would occur even if the speech is of a high quality, whereas visual displays can be sampled at the operator's pace and are permanently displayed. Speech presents a paradox of operation. Participants are required to respond immediately or the message will be lost, but have to wait until the message has ended before responding.

Stokes *et al.* (1990) suggest that speech-based warnings serve as both an attention attractor and as a channel for transmitting information about the failure. Although the synthesized speech used in the investigation was distinctive from human speech, participants in the speech-only condition were significantly slower to accept alarms than participants in the other two conditions, as shown in Table 5.1. Either participants in the speech only condition must have ignored the message, or they failed to realise that the spoken alarm required an immediate manual response. This notion is consistent with Wickens' SCR (Stimulus-Cognitive processing-Response) Compatibility theory of information processing. Simply put, it posits that the input and output to the information processing systems is required to remain in the same modality if performance is to be maximized. Therefore a speech-based alarm would require a vocal, not a manual, response.

In an attempt to understand the nature of alarm handling, a model of alarm initiated activities was proposed (see Stanton *et al.*, 1992). This model suggests six generic states to alarm handling: observe, accept, analyse, investigate, correct and monitor Stanton *et al.* (1992) proposed that the information requirements for each of these stages are likely to be different, and in some cases conflicting. If the requirements can be identified, one could begin to propose where, if at all, speech might be an appropriate medium. Indeed, Stanton (1991) suggested that alarm media have unique intrinsic qualities that might be successfully exploited if used in a sensitive manner, by matching these qualities to the demands of the task. For example, scrolling text favours temporal tasks, plant mimics favour spatial tasks, annunciator panels favour pattern matching tasks, auditory tones favour attraction and simple classification tasks and speech favours semantic classification tasks.In the light of the results reported in this study, it is suggested that synthetic speech is not suitable for semantic classification tasks.

In order to evaluate the appropriateness of the alarm media, it is first necessary to consider the nature of process control tasks and the operators' activities therein. Following a series of experiments examining fault management in process control environments, Moray and Rotenberg (1989) claim that operators might observe abnormal values, but fail to act because they are already busy dealing with another fault and may wish to finish that problem before starting a new one. They term this phenomenon 'cognitive lockup', and note that human operators have a preference for serial, rather than concurrent, fault management. Therefore new alarm information is likely to be ignored until the operator is free to deal with it. If this alarm information is presented through a transitory medium, then it will be lost. This explanation of fault management suggests that synthesized speech-based alarm display systems are not appropriate for process control tasks.

There is some evidence to suggest that irrelevant speech may have adverse effects upon performance (Smith, 1989). The effects appear to occur independently of intensity. Consider the participant working to maintain the process whilst speech alarms are being presented. The alarm information may be described as 'irrelevant' if it does not relate to the particular task in hand, and performance is disrupted.

In conclusion, it is suggested that speech alone as a medium for alarm displays cannot be recommended for tasks where: there is a memory component, there is likely to be some delay before the fault is attended to, there is likely to be more than one alarm presented at a time, and the operator is required to assimilate information from a variety of sources using spatial reference. If speech is to be incorporated into the alarm system for 'process control' tasks, it is recommended that it be paired with other media such as a scrolling text display. However, speech-based alarms might be appropriate for tasks where: an immediate response is required, the 'operator' is away from the control desk, the situation is typically one-alarm to one-event, and fault management is serial in nature.

References

Baber, C., 1991a, *Speech technology in control room systems: a human factors' perspective*. Ellis Horwood: Chichester.

Baber, C., 1991b, Why is speech synthesis inappropriate for control room applications, in E. J. Lovesey, *Contemporary Ergonomics: Ergononomics Design for Performance*. Taylor & Francis: London.

Baber, C., Stanton, N. A. and Stockley, A., 1992, Can speech be used for alarm displays in 'process control' type tasks? *Behaviour & Information Technology*, **11**(4), 216–26.

Edworthy, J., 1990, Auditory warnings for the nineties. *The Occupational Psychologist*, **10**, 23–5.

Moray, N. and Rotenberg, I., 1989, Fault management in process control: eye movements and action. *Ergonomics*, **32**(11), 1319–42.

Patterson, R. D. and Milroy, R., 1980, cited by Edworthy, J., 1990.

Robertson, C. R. and Eberts, R. E., 1987, cited by Baber, C., 1991a.

Smith, A., 1989, A review of the effects of noise on human performance, *Scandinavian Journal of Psychology*, **30**, 185–206.

Stanton, N. A., 1991, Alarm information in fault diagnosis: condition monitoring for fault diagnosis, *Digest*, no. 1991/156, IEE: London.

Stanton, N. A., Booth, R. T. and Stammers, R. B., 1992, Alarms in human supervisory control: a human factors perspective, *International Journal of Computer Integrated Manufacturing*, **5**(2), 81–93.

Stokes, A., Wickens, C. D. and Kite, K., 1990, Display Technology: human factors concepts, *Society of Automotive Engineers Inc.*, Warrendale, PA.

Wickens, C. D., 1984, *Engineering Psychology and Human Performance*, Merrill: Columbus, Ohio.

Part II

6

Speech input

C. Baber

Introduction

In this section we turn our attention to the issues surrounding speech input to computers. As with the previous section we begin with a review/critique of the area, and then consider a range of topics relating to the application of speech input. While we can readily imagine applications of speech output in computers and consumer products, it is often harder to imagine speech input. For this reason there is something of a misconception regarding the utility of speech input. It might surprise readers to learn that research into speech input had been conducted as early as 1911. 'Radio Rex' was a toy dog which jumped out of his kennel when you called his name. In actual fact, he would jump out in response to any loud noise, but he seems to be the earliest manifestation of a voice operated product. The application of speech input really became viable in the early 1970s, with a number of factories in the USA employing speech input on inspection lines. Noyes gives a more detailed account of this in Chapter 21.

Chapters in Section II

The 'Radio Rex' example is instructive because it allows us to introduce a key problem which speech input has had to address: how can speech be sufficiently restricted to permit accurate operation, while being sufficiently flexible so as not to prevent people from using it? Chapter 7, from Bob Damper, considers this question from the point of view of manufacturers' claims regarding the 'naturalness' of speech input. We tend to assume that because speech is the principle medium of communication for human beings, speaking to computers will be inherently 'natural'. Damper debunks this particular myth, and argues that speech input should only be used where it will provide a significant improvement over other media. While one might consider his conclusion that speech will only be useful as a '1 out of *N* selection' tool rather too pessimistic, his arguments are sufficiently convincing to make even the most hardened advocate of speech input pause for thought.

With reference to the industrial application of speech input, we tend to assume that a major benefit to be had from using speech is its ability to support operator mobility; the operator need not be tied to a keyboard, but can be free to wander around the workplace. In the second paper in this section, Dave Usher considers the use of radio microphones for speech input. He suggests that while a number of applications have used radio links, they are not without their problems. In particular, the radio link used in his study appeared to introduce a constant level of spurious noise into the recognition process. This led to the somewhat paradoxical finding that while recognition scores were both higher and more consistent with the radio link than a direct link, the radio link produced more recognition errors.

One of the principle application areas for speech input is in military avionics; the current version of the European Fighter Aircraft will use speech input for a range of control actions. High performance fighter aircraft, flying in excess of Mach and at low level, require a high degree of manual control by pilots. Any input device which permits constant 'hands on throttle and stick' (HOTAS) control will be very useful in this domain. There has been relatively little work on the introduction of speech input to civil aircraft cockpits. In Chapter 9, Alison Starr, drawing upon her experience with introducing speech into civil cockpits, in the FANSTIC project, discusses the potential benefits to be gained from the use of speech in both civil and military cockpits.

When we consider the use of speech input in office applications, by far the most common reaction seems to be a wish for a speech driven typewriter – allowing us simply to dictate complete documents from the comfort of our *chaises-longues*. Alan Newell and his colleagues at Dundee University have been considering the problems associated with 'listening typewriters' for a number of years, and in Chapter 10, by Iain Murray and colleagues, reviews this work. Briefly, the chapter considers approaches to simulating listening typewriters through the use of a 'Wizard of Oz' technique which cleverly employs a palantypist to provide responses to user input. The research has provided a number of interesting conclusions of relevance to speech research in general.

Finally, Chapter 11 is related to the Murray *et al.* paper in that it considers a potential use of speech in office tasks. Philip Tucker and Dylan Jones investigate the use of speech input for document annotation tasks. In particular, they note that documents presented on VDUs take longer to annotate and proof read than hard copy versions of the same documents, and ask whether speech could improve performance of these tasks on a VDU. Their studies suggest that speech may be useful for some, but not all, annotation tasks, and Tucker and Jones argue that speech ought to be used in conjunction with other annotation media.

The papers by Murray *et al.* and Tucker and Jones relate to points made by Damper, concerning the potential use of speech input. One could argue that a number of manufacturers have attempted to force speech input to run before it had mastered crawling, and this seems especially true when we consider office applications. However, when speech input is used appropriately, i.e., in the right manner in the right environment, there is every reason to believe that it can yield performance benefits over conventional media, as shown by Starr.

7

Speech as an interface medium: how can it best be used?

R. I. Damper

Abstract

Intuition, and some influential research in the experimental psychology literature, suggests that speech is the human's most natural communication mode. Based on this, there is a widespread assumption that speech represents the ultimate medium for human–computer interaction; yet speech technology remains little-used in real applications. A possible reason for this is ignorance on the part of systems designers of what the technology can do, or how best to employ it. While the art of designing speech-based systems continues to make steady, if slow, progress, there do seem to remain more fundamental problems. One obvious cause might be limitations in the capabilities of current technology (which we can hope to overcome in time). A further possibility is that speech, while offering some specific advantages, is actually a poor medium for many human–computer interaction tasks. This paper attempts to reach a balanced view of the advantages of speech relative to other interface media and, thus, of the likely rôle of speech in future interactive systems. We focus on input to keep the discussion tractable. Analytical and experimental methods for comparing speech with competitor media are reviewed and discussed.

Introduction

Over recent years, considerable research effort has been directed at realising major advances in the technology of speech communication with computers. This effort has been motivated by the assumption that speech technology offers the key to dramatic improvements in the effectiveness of the human–computer interface. An important but usually implicit, thread is the notion that speech is somehow a universal medium, good for all situations. For instance, Lea (1980) writes:

> . . . you will want to use speech whenever possible because it is the human's most natural communication modality . . .

while Viglioni (1988) says:

> . . . With speech recognition and speech response systems, man can communicate with machines using natural language human terminology . . .

and Lee (1989) states:

> Voice input to computers offers . . . a natural, fast, hands free, eyes free, location free input medium.

While the superiority of speech as an input medium is most often merely assumed, some authorities refer to the experimental psychology literature on problem solving (e.g. Ainsworth, 1988, pp. 3–4). For instance, Chapanis (1975) and his co-workers (Chapanis *et al.*, 1977) have shown a clear advantage to the use of speech in co-operative problem solving between humans in terms of solution speed.

However, there are obvious and marked differences between human–human and human–computer interaction which mean that advantages in the former case do not necessarily transfer to the latter situation. Ainsworth (1988) gives one reason such transfer might occur:

> Presumably as speech is a more natural form of communication it is possible to think and speak simultaneously, whereas complete sentences need to be composed before they can be written or typed.

On the other hand, if speech really carries all the advantages that have been claimed for it, one is entitled to ask why it has not been more widely used in interactive systems.

Why is speech not more widely used?

In spite of many years of optimistic predictions to the contrary, speech remains a very little-used medium of human–computer communication. Certainly, there have been successful applications – notably in hands/eyes busy situations, where mobility or computer access via the telephone network are important, or when the user is unable to operate more conventional input devices because of physical disability. However, these remain small-volume, essentially niche, markets. In many other cases in which speech has been tried, it has been found that conventional input devices serve as well as, and very often better than, speech.

There are perhaps three possible reasons why conventional input devices seem to serve as well as, or better than speech in most circumstances. The first is that successful integration of speech into an interactive system requires a profound understanding of the unique nature of the medium, and the development of new human engineering techniques, which are only now emerging. While there are good reasons for believing this, it does not seem to be the whole of the story. The second possibility is that current technological limitations (which we can realistically hope to overcome in the future) impose a burden on the user which compromises the application. Finally, it may be that speech (far from being 'universal') is actually a rather poor general-purpose interface medium, although it may still offer worthwhile advantages in a restricted set of special circumstances. This last possiblity is not entirely independent of the first: if the advantages of speech are peculiar (i.e. restricted and not obvious), then human engineering techniques will need to be evolved to exploit them.

Newell (1985) gives several reasons why speech might be less than ideal as an interface medium. In his words:

> . . . a major justification for the use of speech has been that it is the 'natural' method of communication for man . . .

and:

> . . . in general, much greater thought must be given . . . than is implicit in justifications of this nature.

An interpretation of Newell's view is that speech input should be assessed rigorously - rather than by asumption. To do this, we need measurable goals.

Human factors goals and methodologies

Shneiderman (1987) has suggested the following human factors goals as suitable for quantifying the efficiency and usability of an interactive system:

- Speed of performance
- Rate of errors
- Subjective satisfaction
- Time to learn
- Retention over time

These, then, are the yardsticks against which any interface, including one based on speech, should be evaluated.

In this chapter, we consider three ways that success in meeting these goals might be predicted, namely:

- by analysis of the input tasks involved;
- by case studies, either 'laboratory-based' or 'application-based'; and
- by so-called Wizard of Oz simulation - see Fraser and Gilbert (1991) for a comprehensive and up-to-date review.

Analysis of input requirements

If speech really is an effective and universal interface medium, then it would be expected to score highly on all the dimensions listed by Shneiderman. If on the other hand, there are particular applications for which speech is appropriate and others for which it is less good (as assessed by the measurable goals), we need some analytical framework in which to assess the likely success of a specific application.

One very useful classification of input (sub)tasks has been put forward by Foley *et al.* (1984). They were primarily concerned with graphics input, but their scheme can be usefully applied to other domains. Foley and his colleagues identify six 'primitive' interaction subtasks:

- selection
- string
- quantify
- orient
- position
- path

The selection function is illustrated, for example, by the common requirement to pick an item from a menu. The string function involves composing a sequence of characters selected from some set as in text composition; hence, it can be viewed as a sequence of selection operations. Quantify calls for the specification of some scalar (uni-dimensional) quantity denoting, for example, the point in some file where editing is to take place. Orient specifies an angular quantity, such as the orientation of a line segment. Position identifies a point in (usually) two-dimensional space - effectively a dyad of quantify operations. Finally, path describes an arbitrary curve in the applications space and can be seen as a sequence of either position or of orient. Each of these 'primitives' can be seen as implying the existence of some abstract device which ideally implements that subtask.

Real, physical devices map onto the abstract devices in vastly differing ways. Thus, whereas the conventional, qwerty keyboard is essentially a string composition device, it can be used to simulate the other abstract devices with greater or lesser facility. For instance, augmented with cursor keys, it can perform the position function. However, a pointing device such as a mouse is a far better realization of the abstract position device. In turn, a data tablet used in conjunction with a stylus is a generally better realization of the path abstract device than is a mouse, even though path can be viewed as a sequence of position, because a stylus has far better handling dynamics than a mouse.

Examination of the way that speech input maps onto the abstract devices gives a useful means of evaluating its strengths and weaknesses. Most obviously, a powerful speech recognizer (used as an automatic dictation machine) would appear to be a very promising text-composition (string) device. For a direct implementation, however, this application requires large-vocabulary capabilities beyond what is currently available commercially. Further, speech input is potentially an excellent means of 1-out-of-N selection, especially for N (i.e. vocabulary size) of the order of hundreds, since this is very close to the recognizer's real mode of operation. By contrast, a keypress device with several hundred keys would be unwieldy in use. On the other hand, it should be readily apparent that speech recognition is a poor match to the requirements of, for instance, quantify and position. The former calls for an essentially analogue, continuous device (whereas recognizers produce discrete output), while position requires spatial, pointing abilities which speech does not really possess.

Given this analysis, speech certainly does not appear to be a universal medium. Rather, it meets the requirements of selection well, with the promise that future technology will deliver good string devices.

Case studies

While the sort of analysis of input requirements described can be extremely helpful, human–computer interaction is a complex subject; it cannot realistically be reduced to the simple procedure of mapping real devices onto abstract devices. Typical of the many other dimensions which need to be considered are the user's physical situation and skills, the cognitive load imposed by the task and safety criticality. As we do not fully understand all the factors impacting on performance, it remains mandatory to carry out task-related case studies. The basic methodology which has evolved is to use the insights offered by such means of analysis as do exist to design a prototype interface which is then evaluated against the measurable goals, listed above.

A number of investigations has attempted to assess the relative merits of speech and keyboard input via case studies. However, according to Simpson *et al.* (1985):

> Research comparing speed and accuracy of voice versus manual keyboard input has produced conflicting results, depending upon the unit of input (alphanumerics or functions) and other task-specific variables.

However, the analytical framework provided by Foley *et al.* (1984) seems to provide an explanation for much of this conflict without ascribing it to imponderable, 'task specific' factors.

Small-cardinality selection

Consider first the issue of 1-out-of-*N* selection. Certainly, the entry of numeric data by keyboard in a simple, primary task (i.e. without concurrent, secondary tasking) is faster and less error prone than entry by speech (Welch, 1977). This is only to be expected as a comparison of speed of keypressing with the time to utter a word makes clear. Hershman and Hillix (1965) found that targeting the finger and depressing a key on a qwerty keyboard took some 200 ms for a practiced typist; the corresponding figure for unskilled typing is some 1000 ms when typing meaningful text (Devoe, 1967). By contrast, it takes some 400–800 ms to utter a typical spoken command with present-day recognizers imposing a further processing delay. In the case of discrete-word input, an additional overhead is introduced by the need for a distinct pause between inputs.

Given these figures, it seems unlikely that spoken entry of primary data of low cardinality (i.e. small *N*) using isolated words could ever be competitive with keying by even a moderately practiced typist, except in special (e.g. hands-busy) situations.

An example of a study contradicting this pessimistic view is that performed by Poock (1980). Twenty-four subjects followed a fixed scenario in which they entered commands (from a set of 90) typical of a naval application, such as 'go to echo' and 'forward message', either by speech or keyboard. The recognizer used was the Threshold Technology T600. Poock claims that speech input is vastly superior to keying in this scenario. The stated findings were that speech input was some 17.5 per cent faster than typing, while typing had enormously more errors (183.2 per cent) than speech command entry.

Crucially, however, the spoken commands were entered as single utterances (speaking e.g. 'go to echo' as a connected phrase) whereas the typed commands had to be keyed character-by-character on a qwerty keyboard. This means that 'go to echo', for instance, would require 10 or 11 separate input (keypress) acts, each of which is time-consuming and error-prone. Yet for the purposes of selecting between 90 actions, there is absolutely no need for keyed commands to be this verbose; one-, two- or three-letter commands should be perfectly adequate - even allowing for them to have some sort of mnemonic content.

Rather than comparing speech and keypressing as input media as such, Poock is in fact comparing:

1. a reasonable interface design in which a single spoken command effects a selection with a much less reasonable design in which an excessive number of keyings is required to select that action, and;
2. the mapping of a real 1-out-of-*N* device (the speech recognizer) to an abstract 1-out-of-*N* selection device, with the mapping of a real string device (the qwerty keyboard) to the same abstract 1-out-of-*N* selection device.

It is hardly surprising that the former mapping is more successful than the latter, regardless of the difference in input media. Indeed, it would be interesting to see how a 90-key pad allowing commands to be selected by a single keystroke - a much closer physical realization to the underlying abstract device than the qwerty keyboard - would have performed.

In an attempt to overcome these methodological objections, we have recently carried out a slightly simplified version of Poock's experiment (Damper and Wood, forthcoming). Twelve subjects entered commands (from a set of 40 rather than 90) from a fixed naval search-and-rescue scenario in two different conditions. In both conditions, the keyed commands were identical; they were acronyms such as 'GTE' for 'go to echo'. In the first condition, spoken commands were 'natural', consisting of the whole phrase (e.g. 'go to echo') as in Poock's study. In the second case, however, they were the spoken equivalent of the keyed acronym (i.e. subjects said 'g t e'), so that the requirement for subjects to recode the command string to an acronym was maintained across input media. The recognizer used (Interstate SYS300) was similar in specification and performance to that employed by Poock.

Results did not differ markedly between these two conditions. Overall, speech input was slightly slower (10.6 per cent) than keyed input but not significantly so. However, using our more reasonable coding of the keyed commands, error rates for keying were very significantly lower (78.3 per cent) than for speech. Thus, contrary to Poock's finding, our study indicates that speech is an inferior medium for the sort of small-cardinality selection which is typical of command and of control applications.

String composition

Recently, it has become possible to perform preliminary, experimental comparisons of speech and keyboard entry of text strings using large-vocabulary recognizers.

Brandetti *et al.* (1988) used the TANGORA recognizer developed by Jelinek and his colleagues at IBM (Jelinek, 1985) in such a study. According to these authors, the prototype for the Italian language 'recognizes in real-time natural language sentences built from a 20 000 word vocabulary'. Subjects entered a 553-word text on two occasions: 'once they used the voice recognition capability of the system, and the other time they used the keyboard only'. (Note the implication, confirmed by a personal communication, that the 'speech' condition is actually speech plus keyboard.) For eight non-typist subjects, it required less time to enter the raw text using speech input than using the keyboard; the relative average figures are 15.5 min for speech and 22.4 min for keyboard. Errors were significantly higher for speech; an average of 7.75 per cent of words in error for speech as against 1.75 per cent for keyboard. The total time to produce corrected text was found to be comparable for the two conditions: 29.7 min for speech and 28.6 min for keyboard. (These figures, averaged for all non-typist subjects, are not given explicitly but have been computed from Brandetti's published data.)

For two professional-typist subjects, total times were 40.0 min for speech as against 22.0 min for keyboard, with errors of 8.8 per cent and 0.5 per cent respectively. Subjectively, it is said that:

> . . . users found more pleasure and satisfaction in the usage of voice rather than keyboard.

Finally, the authors believe that use of a . . .

> . . . voice activated text editor indicated that large-vocabulary speech recognition can offer a very competitive alternative to traditional text entry.

The simulation of future systems

When interactive systems are based on emerging, developing technologies, as in the case of speech, the usefulness of case studies is limited by the performance of available devices. Thus, much human factors work in this area could be criticized as over-concentrating on the shortcomings of currently-available equipment rather than on more fundamental interaction issues, thereby assisting the development of the next generation of interface technologies. According to Simpson *et al.* (1985):

> By simulating speech recognition hardware, various levels of speech-recognition capability can be controlled and evaluated experimentally.

Since such studies use a human operator hidden from the experimental subject to simulate the recognizer, they are generally called Wizard of Oz (WoZ) studies. The hope is that data collected from such simulations will be invaluable in determining the maximally useful properties of future speech systems.

Gould *et al.* (1983) performed an important, early WoZ study in which the capabilities of a simulated speech input device were constrained in various ways. They asked the question: 'would an imperfect listening typewriter be useful for composing letters?' A human typist entering text on a conventional (qwerty) keyboard was the basis of their simulated listening typewriter (SLT). This means that the simulated 'recognition' was not real time. The SLT was compared with baseline performance set by handwriting. The particular imperfections studied were limitations of vocabulary size and the need for artificial pauses between utterances. Thus, the various different versions of SLT used had:

- either 1000-word, 5000-word or unlimited vocabulary;
- either isolated- or connected-word capability.

The vocabulary restriction was simulated by matching the typist's keyboard entries to words stored in a fixed-size dictionary. Words not in the dictionary could be entered by the subjects in a spell mode. Feedback to the subjects used a visual display unit.

Subjects were professionals used to office work, including groups with and without dictation experience. They were required to compose letters by various means: namely, speech input using the SLT (called the 'speech' condition), writing or dictation onto audio tape for later transcription. Note that the form in which the letter was produced is different in the different conditions. Subjects were asked to adopt two separate strategies:

- draft - in which text-entry errors were to be ignored and left to a final editing stage:
- first-final - any errors encountered were to be corrected at the time of entry.

For the writing condition, however, only the first-final strategy was employed.

Results were assessed in terms of speed, subjective preference and the 'quality' of the letters produced as judged by a panel of vetters. Fig. 7.1 shows composition time only (i.e. excluding correction time) for the 1000-word and unlimited vocabulary conditions under both draft and first-final strategies.

Overall, except for small-vocabulary, first-final as shown in Fig. 7.1, use of the SLT was found to be faster than writing but slower than dictation (although the latter does not produce output in 'real time'). As expected, the isolated-word SLT was slower than the connected-word input and speed generally increased with vocabulary size. The subjects mostly preferred the speech condition, presumably because of the combination of reasonable speed and observable results in 'real time'. Interestingly, large vocabulary size appeared slightly more important than connected-word capability. Finally, and not entirely surprisingly, letter quality was more a function of composition time than of means of input. This latter finding argues that, for creative writing at least, speed of input is not a vital issue since the slow process of composition is the limiting factor.

Gould *et al.* were influential in establishing simulation as a legitimate methodology in the study of the human factors of speech input but does have a number of shortcomings. First, the use of a qwerty typist sets an artificial limit on simulated speech-entry speed according to the speed at which this keyboard can be operated. Importantly, it also means that a comparison with (conventional) keyboard entry - a much more serious competitor medium than handwriting, being both faster and yielding 'softcopy' - is not really possible. Second, the SLT does not simulate error patterns (substitutions, rejections, deletions and insertions) at all realistically. In

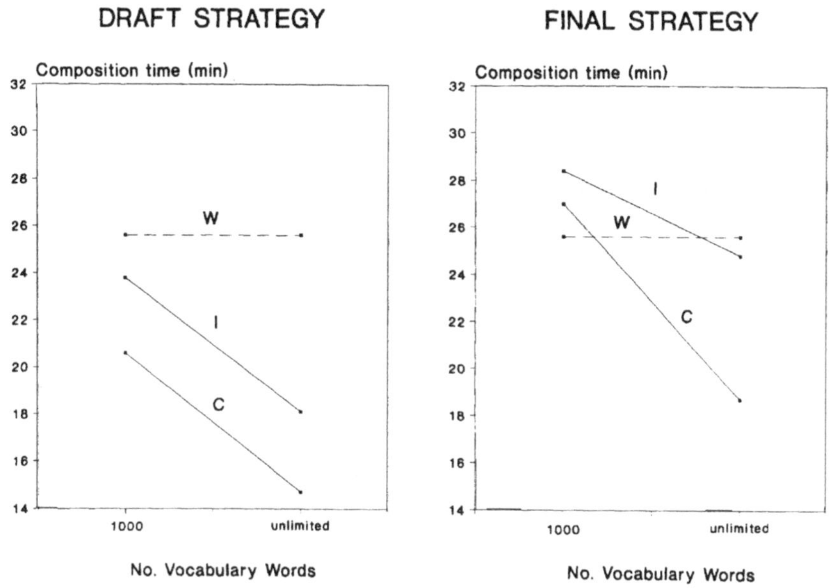

Figure 7.1 Mean composition times for isolated-word speech (**I**), *connected-word speech* (**C**) *and handwriting* (**W**), *after Gould* et al. *(1983).*

fact, all errors were either rejections (out-of-vocabulary utterances or typographical 'errors' on the part of the typist) appearing as XXXX's on the screen, or simulated homophones (the highest-frquency token of a homophone pair was always selected.) Additionally, Simpson *et al.* (1985) implicate 'inconsistent restriction of discrete data entry when the spelling mode was used . . .' as a shortcoming. Finally, post-editing of the draft documents was done by handwriting: thus, the comparison of draft plus edit times with first-final composition times is difficult since it involves a mix of media.

More recently, Newell and his colleagues (e.g. Carter *et al.*, 1988; Murray *et al.*, 1991) have addressed some of the shortcomings of the Gould *et al.* work in their simulation of a speech-driven word processor (SDWP). In particular, they use a Palantype (machine shorthand) keyboard in place of the qwerty keyboard, much reducing the response time of the simulated recognition system. Their studies have addressed the editing of documents by speech by providing speech as the sole input medium.

Input is effectively connected-word with a very large vocabulary of some 13 000 words. Using university students as subjects, composition rates for the SDWP (7.9 words/min) were not as high as those obtained by Gould and his colleagues (11.5 words/min) for the unlimited vocabulary, connected-word SLT using inexperienced dictators and the draft strategy. Also, subjects were less impressed with the SDWP than were Gould's subjects with the SLT, all ranking speech as worse than writing even though the simulation was for full speed, connected speech with a very large vocabulary.

As far as speech editing is concerned, Murray *et al.* (1991) say:

> ... a completely speech-driven word processor ... is an unsuitable system for the fast composition of documents.

Overall, simulation is a powerful way of assessing what would be useful features of future speech systems and, therefore, what ought to be the priorities for development. For instance, the finding of Gould *et al.* (1983) that a listening typewriter restricted to isolated-word input might prove useful, provided it had a large enough vocabulary, would be difficult to obtain by other means.

Concurrent input featuring speech

Broadly, it appears that speech input is not yet competitive with keyed input for primary data entry. However, speech may well hold an advantage when hands and eyes are busy. Thus, it might come into its own in cases of high workload or concurrent tasking when the use of an additional sensory or motor channel becomes important.

As well as examining data-entry performance for simple, random numeric and alphanumeric strings, Welch (1977) also studied performance in a 'complex scenario'. Subjects had to interpret an English language statement relating to simulated flight data, and convert it mentally to a form suitable for entry in restricted fields. In this case, speech entry was faster than keyboard (for inexperienced subjects) and had a comparable 'operational' error rate. However, the speech condition showed a substantially higher error rate before correction. Note the similarity to the results of Brandetti *et al.* (1988), and of Damper and Wood (forthcoming) as described above.

Welch also added a button-pressing secondary task to the primary data-entry task. Although the secondary task did not impact significantly on speed or error rates for the simple data-entry scenario, the complex scenario revealed that the input speed using speech was degraded less severely than that of either keyboard or lightpen entry (Fig. 7.2).

In similar vein, Mountfield and North (1980) studied a dual task in which (simulated) pilots had to keep their aircraft on course using a joystick while, at the same time, having to select radio channels either by speech or keypress. They found that tracking performance using the joystick was degraded very little when radio channels were selected by speech; however, errors were much increased with keyboard selection. Also, radio channel selection errors were higher for the keyboard condition.

It is tempting to explain the speech-input advantage in the dual-tasking studies described above on the basis that the speech channel is additional and does not interfere with the motor channel employed for keypressing in one or other of the tasks. Indeed, we have implicitly taken this to be true in justifying the utility of speech in the hands

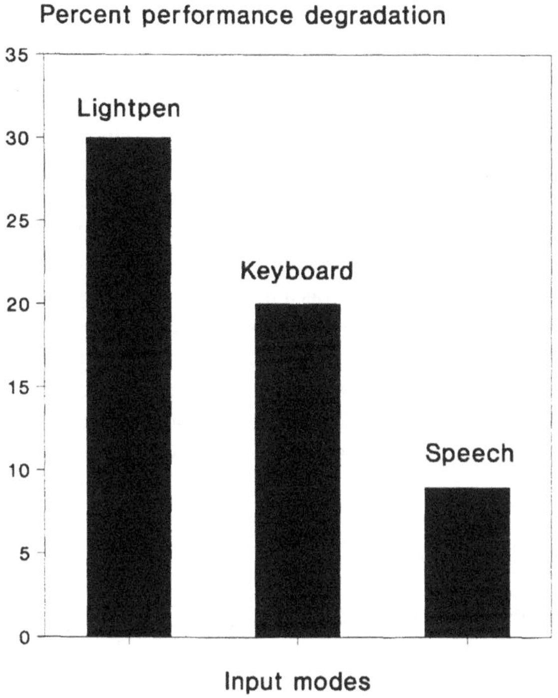

Figure 7.2 Percentage input speed degradation for lightpen, keyboard and voice entry when a secondary task is added to the primary data-entry task (after Welch, 1977)

and eyes busy situation. However, on the basis of experimentation, Berman (1984) cautions:

> the assumption that . . . freeing a particular channel for inputs or responses must necessarily increase the number of tasks or items that can be attended to is not always true.

One relevant finding is the common observation that speech recognition performance under single-task (e.g. list-reading) conditions in the laboratory is always significantly higher than in real or simulated multi-task conditions (e.g. Biermann *et al.*, 1985; Damper *et al.*, 1985). The implication is that task stress competes for information processing resources which would otherwise be allocated to speech production, even when the concurrent task is manual in nature.

Berman (1984) considers a range of models of human information processing and their implications for dual-task interference when speech I/O is used in conjunction with other modalities. His favoured model is that of McLeod (1977), which extends Kahneman's (1973) model by replacing the 'undifferentiated' single reservoir of processing resources by multiple reservoirs of resources. Berman states:

> The implications for speech recognition are that, if it permits the use of a previously less utilized pool of resources, then the total processing resources involved will have effectively increased. However, it also raises the possibility that the change in response modality and,

> potentially, in encoding structures, may act to overload a previously less-burdened pool of resources, and hence cause an . . . increase in task interference.

In a study of the use of speech recognition in television subtitling, Damper, Lambourne and Guy (1985) obtained results which can be interpreted in exactly this way. Speech was compared with keypad for the entry of 'style' parameters (colour, on-screen position etc.), with a keyboard being used for subtitle text entry in both conditions. It was found that speech input of style increased total preparation time by some nine per cent in spite of the fact that the time spent transferring between text and style entry was reduced. Significantly, and counter to our initial hypothesis of diminished task interference, there was a more than offsetting increase in the 'time for apparently unrelated activities as evidenced by longer text-entry, "between-subtitle" and idle times '. This lends weight to Berman's (1984) belief that: 'One should not be misled into anticipating workload reductions by adherence to an inappropriate information processing model' and reinforces the view that task-specific case studies remain essential given the current state of our knowledge in this area.

Summary and prognosis

Simplistic thinking about the virtues of speech as an interface medium is no substitute for rigorous, scientific investigation.

Contrary to apparently widely-held belief, speech is a poor candidate as a universal input medium. It can act as a useful 1-out-of-N selection for reasonably large N, and may offer advantages in situations of dual-tasking, but is otherwise a generally-poorer medium than keypressing at the present time. In the future, however, speech recognition should emerge as a powerful means of fast text (string) entry, but will need to be used in conjunction with a non-speech editing mode.

Wizard of Oz simulation can help define useful characteristics and rôles for future interactive systems featuring speech. The emerging generation of recognizers, such as TANGORA, can also form the basis of useful experimentation. In time, as the technology and human engineering know-how mature, it is to be hoped that the two empirical approaches of simulation and case study will converge.

References

Ainsworth, W. A., 1988, *Speech Recognition by Machine*, Peter Peregrinus, London.

Berman, J. V. F., 1984, Speech technology in a high workload environment, *Proceedings 1st International Conference on Speech Technology*, J. N. Holmes (ed.), 69–76, IFS and Elsevier (North-Holland).

Biermann, A. W., Rodman, R. D., Rubin, D. C. and Heidlage, J. F., 1985, Natural language with discrete speech as a mode for human-to-computer communication, *Communications of the ACM*, **28**, 628–36.

Brandetti, M., D'Orta, P., Ferretti, M. and Scarci, S., 1988, Experiments on the usage of a voice activated text editor, *Proceedings Speech '88, 7th FASE Symposium*, Edinburgh, 1305–10.

Carter, K. E. P., Newell, A. F. and Arnott, J. L., 1988, Studies with a simulated listening typewriter, *Proceedings of Speech '88, 7th FASE Symposium*, Edinburgh, 1289–96.
Chapanis, A., 1975, Interactive human communication, *Scientific American*, **232**, 36–42.
Chapanis, A., Parrish, R. N., Ochsman, R. B. and Weeks, G. D., 1977, Studies in interactive communication: II. The effects of four communication modes on the linguistic performance of teams during cooperative problem solving, *Human Factors*, **19**, 101–26.
Damper, R. I., Lambourne, A. D. and Guy, D. P., 1985, Speech input as an adjunct to keyboard entry in television subtitling, in *Human-Computer Interaction - INTERACT '84*, B. Shackel (ed.), Elsevier (North-Holland), 203–8.
Damper, R. I. and Wood, S. D. (forthcoming), Speech versus keying: a human factors study, *Proceedings of Joint European Speech Communication Association (ESCA) and NATO Research Study Group 10 Workshop on Speech Technology Applications*, Lautrach, Germany, Sept 1993.
Devoe, D. B., 1967, Alternatives to handprinting in the manual entry of data, *IEEE Transactions on Human Factors in Electronics*, **8**, 21–31.
Foley, J. D., Wallace, V. L. and Chan, P., 1984, The human factors of graphics interaction techniques, *IEEE Computer Graphics and Applications*, **4**, 13–48.
Fraser, N. M. and Gilbert, G. N., 1991, Simulating speech systems, *Computer Speech and Language*, **5**, 81–99.
Gould, J. D., Conti, J. and Hovanyecz, T., 1983, Composing letters with a simulated listening typewriter, *Communications of the ACM*, **26**, 295–308.
Hershman, R. L. and Hillix, W. A., 1965, Data processing in typing; typing rate as a function of kind of material and amount exposed, *Human Factors*, **7**, 483–92.
Jelinek, F., 1985, The development of an experimental discrete dictation recognizer, *Proceedings of the IEEE*, **73**, 1616–24.
Kahneman, D., 1973, *Attention and Effort*, Prentice-Hall, Englewood Cliffs, NJ.
Lea, W. A., 1980, The value of speech recognition systems, in *Trends in Speech Recognition*, W. A. Lea (ed.), Prentice-Hall, Englewood Cliffs, NJ, 3–18.
Lee, K-F., 1989, *Automatic Speech Recognition: the Development of the SPHINX System*, Kluwer, Dordrecht.
McLeod, P., 1977, A dual task response modality effect: support for multi-processor models of attention, *Quarterly Journal of Experimental Psychology*, 185–9.
Mountfield, S. J. and North, R. A., 1980, Voice entry for reducing pilot work-load, *Proceedings of the Human Factors Society*, 185–9.
Murray, I. R., Arnott, J. L. and Newell, A. F., 1991, A comparison of document composition using a listening typewriter and conventional office systems, *Proceedings of Eurospeech '91*, Genova, Italy, 65–68.
Newell, A. F., 1985, Speech – the natural modality for man-machine interaction?, in *Human-Computer Interaction - INTERACT '84*, B. Shackel (ed.), Elsevier: North-Holland, 231–5.
Poock, G. K., 1980, Experiments with voice input for command and control, *Naval Postgraduate School Report, NPS55-80-016*, Monterey, CA.
Shneiderman, B., 1987, *Designing the User Interface: Strategies for Effective Human–Computer Interaction*, Addison-Wesley, Reading, MA.
Simpson, C. A., McCauley, M. E., Roland, E. F., Ruth, J. C. and Wiliges, B. H., 1985, Systems design for speech recognition and generation, *Human Factors*, **27**, 115–41.
Viglioni, S. S., 1988, Voice input systems, in *Input Devices*, S. Sherr (ed.), Academic, San Diego, CA, 271–96.
Welch, J. R., 1977, Automated data entry analysis, *Rome Air Development Center Report, RADC TR-77-306*, Griffiss Air Force Base, NY.

8

Automatic speech recognition and mobile radio

David Usher

Abstract

A radio link from a mobile speaker to a speech recognizer is studied in the context of a task drawn from an industrial context - Telecommand. It was found that the task could be successfully carried out using the radio link but it took longer and required more words than when a direct microphone connection was used.

However, it is concluded that the freedom of movement offered by the radio link might compensate for the reduction in recognition success. A speech recognition system could be used in this way if speaker mobility were paramount in the application.

Introduction

The use of automatic speech recognition (ASR) for the human–machine interface has been increasing in recent years. The performances of speech recognizers now on the market, such as those developed by Marconi, Votan and Kurzweil, have been found adequate for industrial applications ranging from the aircraft cockpit to parcel sorting. However, the use of ASR raises many human factors issues (Bussak, 1983) which have been less well studied, despite their crucial importance to the success of the human–machine interface as a whole.

One such issue is user mobility. If the speaker were linked to the ASR device by radio rather than a microphone cable, he or she would be able to move around a large workspace whilst continuing to interact with the plant.

This would make ASR available to a larger variety of tasks and users. Candidate tasks might include the inspection of machines or properties, stock taking, surveying or operating a wall mounted control panel. The ability of the speaker to record data in machine-readable form without use of hands and without any other physical constraints can be seen as invaluable in many situations.

This paper describes a study of the use of a radio link for ASR, in the performance of a task taken from an industrial context, and compares it with a direct microphone connection.

Hardware

Speech recognizer

The speech recognizer used for the study was the SR128 manufactured by Marconi Speech and Information systems at Portsmouth. This is a speaker-dependent device, requiring a version of each word in the vocabulary to be 'enrolled' by the speaker before recognition can begin. At enrolment, the background noise is first sampled and then the user is prompted with the text of each word in the task vocabulary and asked to speak it. The speech signal is analysed every 20 ms and the powers contained within each of 19 different frequency bands are stored. These sets of data (templates) can be read back from the device and displayed graphically, using purpose written software (Usher, 1986).

In recognition mode, the SR128 applies the same digitizing algorithm to the incoming speech and compares the results with the templates. It reports to the host computer the number of the most similar template and a 'score' in the range 0 to 255 quantifying the match. Speech samples of up to 4 s duration may be searched for templates in this way.

The recognizer also has a syntax facility which allows the search to be constrained to a particular subset of templates determined by the recognition history. However, whereas this feature works well in the absence of recognition errors, it can cause confusion and frustration if ever a subset is activated through misrecognition. For this reason, the entire vocabulary for the study was stored on the same syntax node.

Radio

The radio equipment used was the CGX0520S mobile system manufactured by Maxon Europe Ltd of Watford, of the type used by the police. It operates at a frequency of 462.457 MHz FM and a power level of 2 W. There was a handset (Fig. 8.1) which could be carried in a leather case attached to the user's belt, a headset incorporating a noise-cancelling microphone, a VOX unit and a base station. The output of the base station was matched electrically to the characteristics of the microphone input of the SR128.

The purpose of the VOX unit is to open the radio channel automatically when speech is detected. Early trials showed this to be unsuccessful in this application because of the 2 s lag before transmission started. The alternative 'press-to-talk' mode was used, since continuous radio emission is permitted by neither battery technology nor radio licence.

It was also quickly found that the 'squelch' – a burst of broadband noise on termination of transmission – could not be tolerated because the recognizer attempted to match it to the templates (despite the 'wildcard' facility) and produced many insertion errors. The suppliers of the base station (Tranex of Rothwell) modified it and successfully eliminated the problem.

Figure 8.1 The portable radio equipment used in the study

Computer

The software environment was a specialized process-control language called Swepspeed running on a DEC 11/23+. The graphics were generated on a SBD-B board (resolution 768*574) from Dowty Computer Graphics and displayed on a 14″ diagonal touch-screen from Mellordata. Some of the software was created for an earlier demonstration of ASR at Durley Park, Grid Control Centre, Bristol (Usher and Baber, 1989).

Task domain

The task domain chosen for the study was Telecommand – the control of electricity substation circuit breakers from a Grid Control Centre. This choice was made because of the potential of ASR to reduce the workload of system operation staff. Telecommands are couched in a five part format, strictly adhered to, as follows:

AT [location] [voltage] [action] [breaker name]

For example, to open the 400 kV breaker B594 at a substation at Handsworth, the telecommand would be

AT *HANDSWORTH 400KV OPEN B594*

The grid control staff speak the telecommands over the telephone to personnel at the substation, who then repeat them for confirmation, and then both speakers make a written log of the transaction. Clearly, this is a domain in which ASR could bring appreciable benefits. If the data were once captured in a computer, the logging, display and transmission of the telecommand could be automated.

The regime

A demonstration package of ASR for Grid Control Rooms was used for the study. Simulated data and displays could be brought up on the touch-screen by spoken command. The 'overview' display shown in Fig. 8.2 was a fictional grid network diagram containing 14 circuit breakers at nine different substations and three different voltages. As each template was recognized, its associated text appeared within a feedback window in a slot determined by its rôle in the telecommand.

As has been mentioned, the SR128 internal syntax facility was eschewed. This allowed misrecognitions to be corrected simply by repeating the word, a method preferable to entering an error-correction dialogue (Baber *et al.*, 1990). The enrolment screen could be regained by touching a key, so that persistent cases of misrecognition could be addressed.

When a complete telecommand was displayed in the feedback window, the application software validated it for consistency with the system diagram. If it was found meaningful, a touch-sensitive key labelled 'SEND' appeared in the bottom right-hand corner of the screen. On touching this, an acknowledgement was shown, the feedback window was cleared and the mimic was updated to reflect the new status

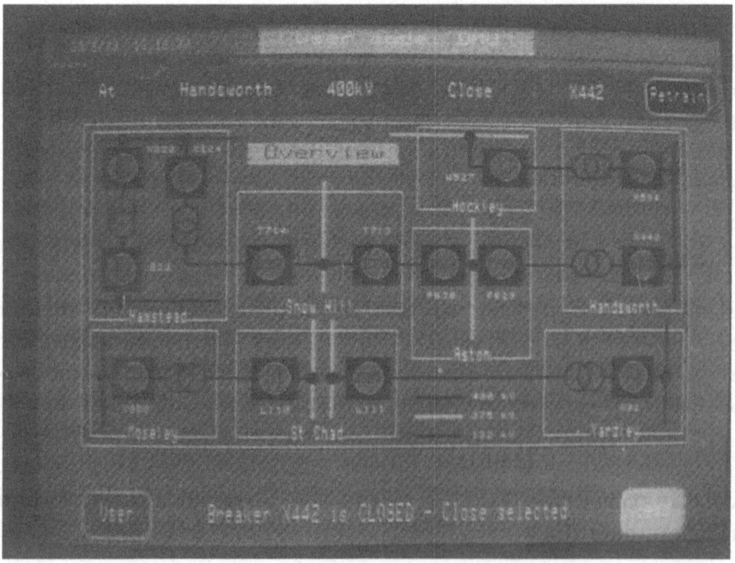

Figure 8.2 The Telecommand display, with a completed command in the feedback window

of the breakers. This process took about 3 s, during which the recognizer was still live, allowing the user to begin setting up the next telecommand. The issues surrounding the use of touch for confirmation are discussed on page 80.

The task

The task was to open all the breakers shown on the display and close them again, which requires 28 telecommands to be sent. It was performed twice with the radio link and twice with the direct microphone connection. The gain of the input stages was adjusted in both cases to produce a minimal background noise mask without speech causing overload.

The vocabulary consisted of 32 words, some of which were irrelevant to the task, such as those for changing the display. The extra words provided additional realism, since an industrial system would not be constrained to a single function.

The entire vocabulary was enrolled immediately prior to the performance of each task, using the relevant microphone interface. The speaker was the author, who can be classed as an experienced user of ASR systems. No re-enrolment of any of the templates was found necessary during any of the repetitions of the task.

The words recognized during the generation of each telecommand, their scores and time tags (to the nearest 0.5 s) were stored on disk when the Send key was touched.

Results

The difficulty of parameterizing the effectiveness of a speech interface is well known (Baber *et al.*, 1991). No absolute measurement parameter has evolved, since the various criteria such as vocabulary size, accuracy, speaker-independence, and continuity of speech may be ranked in a different order of importance for each application. The conclusion is drawn that only task-specific assessment criteria can be regarded as meaningful.

Figure 8.3 shows the times taken to complete all 56 telecommands with the two forms of link. As both of these distributions are markedly skewed, their means and deviations are less than meaningful. However, it can easily be seen that the majority of telecommands issued by radio were generated in a time of between 3 and 7 seconds, compared with under 4 s for the direct link. Thus the radio link is found to slow the execution of the task.

An interrelated measure is the number of utterances required to generate each telecommand. The effect of the radio link on this parameter is shown by comparing the histograms in Fig. 8.4. The direct link data have a very pronounced peak at five utterances (the number of parts of a telecommand) whereas in the case of the radio link there is a less well defined peak at seven utterances.

The recognition scores recorded during the performance of the tasks were normalized by the recognizer itself (so that words of different length might be

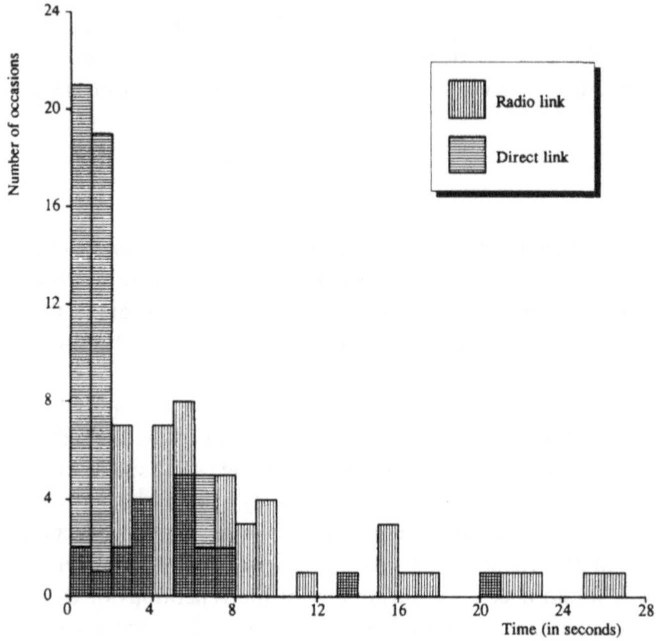

Figure 8.3 Time data

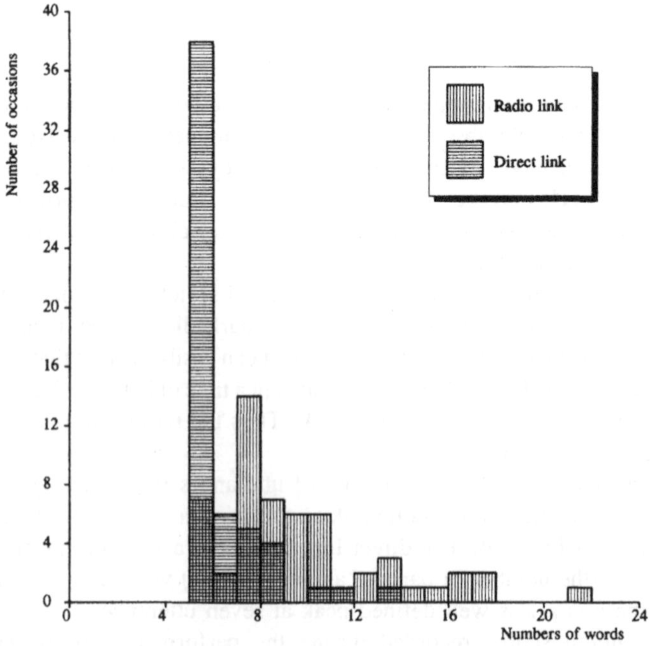

Figure 8.4 Word number data

meaningfully compared) and reversed in sense by the application software (in order that high scores might indicate good matches, as is more natural). Figure 8.5 shows the scores recorded during the use of each form of link. It can be seen from the figures that although the distributions are again skewed, the radio link scores are generally higher. This result is in stark contrast to those obtained from the task-related measures - the telecommand generation times and the utterance numbers. The radio link appears to impose more consistency between utterance and template.

Discussion

Results

Although the radio link has been shown to reduce the effectiveness of the speech interface, it was nevertheless quite possible to issue telecommands in this way. The radio equipment was of ordinary quality, not optimized for the purpose, and the ability of the speaker to adapt and compensate for the perceived failures of the system would be likely in time to reduce the difference between the media - although the ergonomist should be wary of requiring such adaptation of users.

The use of mobile radio for ASR cannot be ruled out by its relatively poor performance compared with a direct microphone link if the application requires the full mobility of the speaker.

The perverse result obtained from studying the scores recorded from the two media

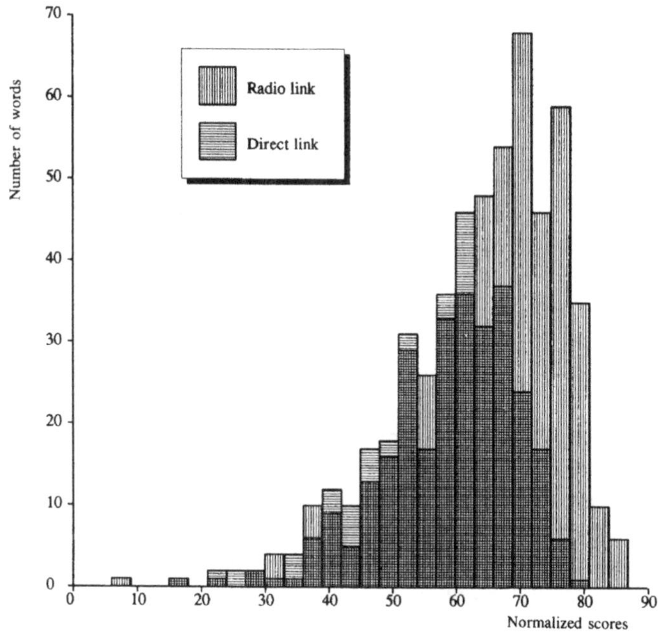

Figure 8.5 Recognition score data

might be due to the static interference upon which the radio signal was superimposed. In the presence of a spectral feature common to all incoming speech, the templates will be more similar (one to another) than otherwise, and once digitized the candidate utterance will match any of them more closely. The effect of audio bandwidth limiting is analogous – the signals contained in frequency channels falling outside the speech band are likely to be similar from one utterance to the next. Therefore, insofar as the overall recognition score is an amalgamation of sub-scores derived from each channel, the effect can be seen as giving the scores a higher starting base, whilst having a deleterious effect upon the recognition accuracy. This result reinforces doubt over whether such recognition scores can provide meaningful information on the performance of ASR.

Human factors

Apart from the results described above, the study provided an opportunity to assess, in a more general sense, the human factors of mobile radio for an industrial application of speech recognition.

Weight

The radio handset was very robust in its design and seemed very reliable, but it is somewhat irksome to carry a package weighing over 1 kg throughout the working day. However, the mobility offered might allow the user to carry out the task from a sitting position or while consulting documents remote from the workstation, which would bring about a reduction in physical effort.

The handset contained batteries which needed regular charging, but did not indicate to the user its state of charge. This is a design fault which should be easily rectifiable, but such technology is perhaps underdeveloped.

Press-to-talk switch

The press-to-talk switch might have been easier to operate. It was small and, being clipped to the user's clothing, rather difficult to locate. Notwithstanding the drain on the batteries, a latching switch would have been beneficial.

Confirmation

A further issue concerns the confirmation of the command. The error inherent in voice recognition means that if it were used both for generating commands for a system and for entering them, it would be possible to enter an incorrect command unintentionally and (worse) unknowingly. For this reason a 'send' key was used in the telecommand demonstration instead of a spoken confirmation.

In order to maintain full mobility, an application would either require an additional radio channel dedicated to the confirmation of the command, or a method of injecting an unmistakable (non-speech) signal into the speech channel. The user would confirm the candidate command by pressing a push-button perhaps mounted on the radio's carrying case.

Vocabulary

It should be noted that the correspondence between text and speech is entirely determined at enrolment. When prompted with the text '132 kV', a user might say *One three two kilovolts*, *A hundred and thirty two kay vee*, *One three two* or even *low tension*. This is one of the advantages of ASR: the user can customize the task vocabulary at will, as long as he or she can remember the choices made.

In this context a radio link possesses a unique advantage. The fact that the radio is worn on the person effectively ties the speaker to the set of templates. The importance of using the correct set of templates cannot be over-emphasized: it is for this reason that the Telecommand screens show a user code so prominently. Where an ordinary desk-mounted microphone is used for ASR, the system must always provide the facility for a new speaker to 'sign on', either by enrolling or by loading pre-trained templates. But there is always the possibility that the speaker might fail to do so, and attempt to use the wrong template set. The likelihood of this occurrence is minimized by the use of mobile radio, since signing on can be seen as an integral part of putting on the equipment when beginning work.

Feedback

It could be argued that the use of a radio link for speech recognition is fatuous since the need to display the recognized words for confirmation confines the speaker to within viewing distance of a screen. However, there are feedback media to which this criticism does not apply, such as large vacuum fluorescent displays, projected video images or even synthesized speech, since mobile radio channels are generally two-way.

Rather than using text messages, the recognized words might be represented symbolically (Baber *et al.*, 1992). This has the advantage of not being confined to computer generated graphics, since a hard-wired mimic could indicate which plant item has been addressed (by flashing a light behind it, for example) and also the state into which the telecommand would put it. This would make it feasible for an operator of complex plant shown on a large wall-mimic to issue commands by voice from any point in the control room and obtain feedback from the mimic itself.

However, the use of symbolic voice recognition feedback for the operation of plant though a wall mimic would effectively confine the interface to a single user. This is a human factors constraint rather than a technical one, because although it would be technically possible for any number of ASR channels to terminate in common at the mimic, such a system would be hopelessly confusing as plant items would be addressed and actions initiated unexpectedly from the point of view of each user.

These considerations point towards the use of vacuum fluorescent text displays for voice-operated wall mimics, with each user allocated one line for his or her speech recognition feedback. Displays of this type are commercially available in a large range of colours and text sizes.

In such a control-room, each of the personnel would be able to read the others' potential commands, as they would be displayed adjacent to their own. This can be seen as advantageous since it would give each worker an overview of the plant status and forestall potential conflicts in its operation. It would be an example of improved human-machine communication bringing with it an improvement in communication between humans.

Conclusions

A radio link from mobile speaker to a speech recognizer has been studied in a task drawn from an industrial context - Telecommand from the Grid Control domain. It was found that Telecommand could be carried out through a radio link, but not without a marked deterioration in performance. It took longer and required more words than when a direct microphone connection was used.

The automatic radio channel activation device (VOX unit) was found unusable because of its slow response to speech. Also, as in other ASR work, the on-board syntax feature of the recognizer was not found beneficial. The results call into question the usefulness of the score attached by the SR128 and other ASR devices to all the recognition report, since this parameter showed improvement when the radio link was used - the opposite effect from those shown by the task-related performance measures.

It is also concluded that the reduction of recognition success associated with the radio link might be mitigated by the freedom of movement it offers the user. A speech recognition system operating through a radio link can be considered usable if speaker mobility is paramount in the application.

Acknowledgement

Thanks are due to Mike Whileman of Tranex Telecommunications for his willing assistance.

References

Baber, C., Stammers, R. B. and Usher, D. M., 1990, Error correction requirements in Automatic Speech recognition, *Contemporary Ergonomics*, 454, Taylor & Francis.

Baber, C., Stammers, R. B. and Usher, D. M., 1991, Assessing automatic speech recognition systems.

Baber, C., Usher, D. M., Stammers, R. B. and Taylor, R. G., 1992, Feedback requirements for Automatic Speech Recognition in process control rooms, *International Journal of Man-Machine Systems*, **37**, 703–19.

Bussak, G., 1983, Voice leaders speak out, *Speech Technology*, **1**(4), 55–68.

Usher, D. M., 1986, A software package for the SR128 speech recognizer, SWR/SSD/0782/N/86, CEGB.

Usher, D. M. and Baber, C., 1989, Automatic speech recognition in Grid Control Rooms, TD/STM/89/10024/N (part 1), TD/STM/89/10031/N (part 2), CEGB.

9

Is control by voice the right answer for the avionics environment?

A. F. Cresswell Starr

Introduction

Since the 1950s speech recognition techniques have been explored with a view to their real-world application in the control of systems. Due to the demands of the airborne environment and nature of the benefits relating to the technology of speech recognition, aviation has been one of the areas seen as having potential for reaping benefits from such an implementation. This has been especially so in the world of military fast jet operations and helicopters, and this application area has been a driving force in the development of the technology, as much of the research for aviation has been related to the military world.

The possible suitability of speech technology to the aviation environment comes primarily from the potential benefits afforded by 'hands and eyes free' systems management, control and data input. There is a requirement, particularly on military fast jet pilots, to spend as much time as possible with their eyes looking out of the cockpit, as their visual attention is best utilized to observe the outside terrain and enemy rather than looking down inside the cockpit to read instruments, operate controls, enter data, etc. This is especially true when the pilot is controlling the aircraft in manoeuvres close to the ground at high speed over enemy lines. However, in addition to flying skills required for ground and enemy avoidance, systems management must also be achieved for mission success. In addition, and more devastating than mission failure, accidents can arise from poor or inefficient interface design. In the year from October 1987, NATO lost 100 jet aircraft of which two-thirds were attributable to pilot error (Gorden, 1989). The contributing factors were stated as overload, disorientation or distraction which keeps the pilot with his/her head in the cockpit. Analyses of accidents over a seven year period of F-16 and F-15 military jets operations indicated that a significant number of these accidents could have been avoided if speech technology had been implemented to allow maximum 'head up' and 'eyes-out' time, preventing disorientation and distraction during the most critical and difficult phases of flight (Gorden 1989). In addition, technology and systems onboard the aircraft have, over the years, increased in complexity and tasks the pilot must complete, thus ever increasing pilot workload. This is reflected in research based on simulated combat exercises where two-thirds of the flight into terrain accidents were due to the pilot becoming saturated in complex cockpits and mission stress (Gorden, 1989). The question is how many accidents and incidents a year would be prevented in the real world with the application of speech technology. Historically, switches for controlling primary systems have been added to the flight controls of aircraft in order to try to overcome the problems requiring system interaction with the pilot matching head up and eyes out. However, the memory requirement for switch functionality only adds to the pilot's already high workload. Similarly, the military helicopter environment has been seen as a primary application domain for speech technology. Besides the need to keep eyes out of the cockpit as much as possible, the flight controls for helicopters often require both hands constantly on the flight controls, particularly during take-off, sustained manoeuvres close to the ground, approach to landing and hover (Coler, 1984). This makes any management of, and interaction with the aircraft system difficult and potentially dangerous, if not impossible. Speech technology may well offer a control technique which afford interaction with systems while maintaining eyes-out and hands on flight controls.

The world of civil aviation is markedly different from that of the military fast jet and the design issues are therefore dissimilar. While the basic skills of flying and navigation are the same, there are many differences. The pattern of the workload and the detailed nature of the tasks to be carried out are not the same: busy phases of flight such as landing and take off are contrasted by relatively quiet periods such as cruise (especially on long haul flights) in civil aviation. In contrast, the middle of a military jet flight is likely to involve most activity, when the tasks associated when the mission need to be completed, such as reconnaissance, target acquisition or weapon release, in addition to the high risk flying task involved. This requires multi-tasking from the pilot. Overall goals in the civil domain relate to passenger safety, comfort, and operational efficiency, i.e. arrival as specified by the timetable (correct place and on time!), whereas military mission success would be judged in terms of information returned, targets attacked and successful return to base, etc. Hence the drive to solve the problem of efficient systems' management has not been so strong in civil aviation.

The application of any equipment to the avionic domain has to be considered in terms of cost as well as desirability, efficiency and feasibility. The speech recognizers robust enough to cope with the airborne environment and design standards have had a very high cost associated with them. Therefore, the cost of speech technology was generally thought to outweigh the potential benefits of its application to civil aircraft. However, technological advances in speech technology have raised the profile of speech recognition drawing attention to its potential benefits on the flight deck. Air traffic world-wide is increasing and hence there are more aircraft at any one time in the terminal airspace associated with any large international airport. Hence, there is a need for the pilot to exercise more external vigilance. In addition, the information required and available on modern flight decks is increasing and hence the number of systems for the pilot to interact with is growing (Warner and Harris, 1984; White, 1982). On military flight decks, the problem is enhanced by the shrinking size of the cockpit of fighter aircraft and therefore of the space available for controls and displays (White, 1982). The complexity of system tasks has also increased and likewise, the interaction required between the pilot and the systems onboard. These systems and the information within them have to be readily available to the crew, if they are to afford any benefits. Hence, the mechanisms for systems' interaction require review. Speech technology is one of the technologies along with touch, gesture and others now being considered for application on future civil flight decks.

This chapter overviews the human factors issues specific to the application of speech technology to aircraft and briefly discusses the future possibilities for its development into fully operational applications.

Potential applications

In order to consider the application of speech technology to the avionics environment, the potential benefits need to be clearly understood. These are generally considered to be:

- hands and eyes free operation of system(s);
- a reduction in workload, particularly in systems management, hence allowing the pilot to pay more attention to the flying tasks;
- easy of operation in the workplace.

In addition, it appears to offer (Fanstic Task, Synthesis Report, 1992);

- increased control capabilities;
- simplification of the control panel.

Some initial research has indicated that speech technology may well reap benefits when applied to some avionic tasks. Gorden (1989) stated that a voice system would be a tremendous benefit to single-seat aircraft where pilot saturation and distraction in the complicated and complex cockpit are a real problem. In addition to reduced performance, accidents may arise from overloading the pilot. Research on the American F-16 programme investigating the use of speech recognition in low level, fast jet operations, compared missions with and without the use of speech technology in systems control. Pilot comments included the following:

- Speech's real advantage comes when you are real busy;
- Voice didn't change the way I managed the aircraft but it was safer and easier.

In another study, helicopter pilots (Coler, 1984) stated that speech recognition in the cockpit was easy to use, and that there was no distraction to the input/command task when the pilot was also monitoring flight instruments or looking outside for air traffic. With the development of technology in both military and civil aircraft, from single displays, knobs and switches for single functions to multi-function screens and control device requirements, reflect an increasing need for a flexible input device

(Taylor, 1986). As cockpits become increasingly complex the role for speech potentially enlarges. The addition of more systems, and the need to access an increasing amount of data and control significantly more functions means that an interface device which has the flexibility to control and interact in an adaptable manner is required.

One of the major benefits of speech recognition is that a single word or simple phrase can be used to replace a complex key structure or chunk of data, and this has led to the suggestion that speech is an obvious input technology for the selection of systems (FANSTIC Task 4.0, Final Report 1992; White, 1982). Alpha/number keys have been introduced into aircraft to enable the selection and entry of complex data and commands such as navigation points within the computerized flight management systems or military equivalents. In order to do this the pilot tends to need to spend long periods head and eyes down, and replacing a key-based input system with voice would allow this procedure to be carried out head up. It could also simplify the dialogue interaction required with the use of simple commands rather than complex digit strings. For example, during the course of a flight (civil or military) the pilot is required to tune the radio frequency in order to receive information about other air traffic, etc. These radio frequencies consist of a series of numbers, which the pilot currently changes when retuning. In a speech system, the selection of a channel and/or radio type could be replaced by a single command, as the speech recognizer could be preprogrammed to 'know' the radio frequencies associated with various names, e.g. Heathrow Tower.

The number of possible applications within the flight deck are extensive. White *et al.* (1984) working for the aircraft manufacturers Boeing, carried out some research to establish where the greatest benefits from speech recognition may be found. They categorized potential applications into five groups: programming; interrogation; data entry; switch and mode selection; and continuous/time-critical action control. The operational requirements of each task were then considered, and these included the active vocabulary required, the acceptable minimum recognition accuracy, how often the task was performed, and whether isolated recognition would be acceptable, or if connected speech recognition was required. This was followed by an evaluation, based on a set of rating scales, of the possible application against a number of pay-off criteria to establish the potential benefits of using voice to control tasks in each category. The rating scales used were technical feasibility, advantage to the pilot, the time and accuracy requirements associated with the task, suitability of hardware for the task and overall desirability of the application. The evaluation was carried out by a team of seven whose members included pilots, flight-deck researchers and human factors specialists. The results indicated that programming (e.g. entering data to the navigation system, tuning radios by identification, setting up the landing system) and interrogation tasks (e.g. requesting system status information, operating a checklist) were considered to be the most likely to bring benefits to the pilot if controlled by voice. This was especially the case for the application of speech recognition to complex systems which involve selection from menus in their interface design. Data entry tasks were also considered by the evaluation team to show promise for speech input. The research also indicated that speech could be easily used for the selection of modes and switch settings, but this was not considered to provide the optimum use for voice. The application of speech input to continuous/time critical actions, (e.g. volume control on communications equipment, and selection of positions of flaps, speed breaks and trims) was not considered desirable.

Device selection

Speech recognition is one of the potential interface technologies being considered for application to the new and complex data-rich multi-function systems being introduced on the flight deck. Some research work has been undertaken to investigate the relative merits of different devices. For example, White and Beckett (1983) compared the use of various input devices for air crew tasks, including data entry, in a simulated military airborne environment. The technologies compared were keyboard, touch and voice. The study found that the use of voice reduced 'the spatial and temporal distraction of the pilot from the primary task of flying the aircraft', and this meant that there was more time for eyes-out flying. Work at NASA Ames (Warner and Harris, 1984) compared speech recognition and keyboards in simulated helicopter missions with various noise, turbulence and vibration conditions. They found that both input devices could achieve 99 per cent performance under most conditions, with speech input falling to 98 per cent under severe vibration. However, under all conditions input by voice was found to cause less disruption to a concurrent tracking task when compared to tracking task performance with concurrent key-based inputs. Further work by the same group showed that as the complexity of the input task increases the disruption to secondary tasks also increases when keying is used, but not with speech recognition input. Research into advanced crew stations design at the USA Naval Air Development Centre found similar results to those generated by NASA Ames (Warner and Harris, 1984). Dual task performance using voice when compared to conventional aircraft keyboards for control of displays, indicated that the latter caused more interference with the flying task. The study concluded that workload conflicts in dual task situations could be overcome by the use of voice for the control of selected tasks. A series of experiences carried out by Laycock and Peckham (1980) concluded that input via speech was faster and more accurate for data entry than if the task was carried out via a keyboard. In addition performance on a concurrent flying task suffered less degradation when speech was used rather than a keyboard, and that this was especially true for high workload conditions. Similar results were obtained by Mountford and North (1980).

In Enterkin's (1991) paper, which includes a brief review of research in this field, the following benefits have been shown experimentally to be attributed to speech input:

- it decreases input times;
- it increases accuracy in the input of data;
- it improves the attention of the user to concurrent tasks;
- it improves head up attention for concurrent flying tasks.

In line with this, research by Coler (1984) discovered that voice may provide faster, more accurate command input than keying.

The use of speech recognition on the civil flight deck has continued, however, to be controversial, especially with new emerging technologies, such as gesture, touch, etc. bringing with them their own advantages. Little research, directly related to the civil flight deck, appears to have been undertaken. However, the following is a report of one study which took place in the late 1980s and early 1990s, under a European collaborative programme called FANSTIC (Future Air Traffic Control, new Systems and Technology Impacts on Cockpits) (FANSTIC: Final Report and Synthesis Report, 1992). Under this programme, an investigation was carried out by the relative merits of speech recognition, touch and conventional (push button/key) interfaces found on most aircraft flying today (Starr *et al.*, 1992). It involved pilots flying a series of short circuits around Heathrow Airport. The application used as a test bed for the study was the control of two multifunction systems displays mounted in the centre of the control panel of a civil flight deck simulator. These displays presented systems status and checklist information on demand from the pilot. Qualified pilots performed three trial flights and during each flight they used a different control technology for interaction with the system displays. In addition, during each flight variations were made to two types of workload: vocal (other than speech recognition) workload, by forcing the pilot to interact more or less often with air traffic control, and the flying task, by switching the autopilot in and out of the control loop. The pilots were also forced by an experimenter to interact with the system displays at an abnormally high rate to ensure the system was fully exercised. However, the operations carried out by the subjects were operationally feasible. At the end of each flight subjective measures and opinions were collected. Results showed that it was generally felt that speech recognition could be beneficial to the operation of flight deck interfaces, and that it would be acceptable to pilots.

The results of the FANSTIC studies indicate that speech appears to afford benefits in terms of primary and secondary task performance, and therefore from a user performance viewpoint it should be considered further in applications for aircraft.

Dialogue design issues

The interaction that takes place between the user and the system composes a dialogue. A dialogue is made up of various elements including vocabulary items, dialogue structure or syntax, feedback, error correction, etc. The dialogue between the user and the system can take a number of different forms. Initial designs of the speech interface in aviation have been based on isolated word recognition and a menu or tree interaction structure. This type of dialogue means that only a few words are 'active' at one time for recognition, and hence the possibility of recognition errors is reduced. Recognition of one word moves the interaction forward to a new set of words for recognition until a command or complete piece of data has been input. However, this means the interaction is generally stilted and unnatural, and it requires the pilot to remember a very fixed dialogue, thus adding to their workload.

As the technology has developed, and through the introduction of knowledge and artificial intelligence techniques to its design, the dialogue produced can be more flexible. The interaction is moving from recognition towards speech understanding. The speech recognition system designed by the French avionics manufacturer Sextant for the FANSTIC programme was a part step towards a 'natural language approach', as in order to introduce flexibility into the dialogue, the user could use synonyms and invert/alter the order of words. The system also allowed the use of different syntaxes for the entry of numerical values. Historically, a three-figure value sent to a recognizer would consist of a digit, followed by a digit, followed by a digit. For speed, three syntaxes were feasible:

- *digit*, followed by *digit*, followed by *digit*, e.g. two, one, five;
- *digit*, followed by *0/99*, where 0/99 represents the numbers between 0 and 99, e.g. two, fifteen;
- *digit*, followed by *hundred* (or *hundred-and*), followed by *9/99*, e.g. two hundred and fifteen.

The recognizer achieves this by having some 'concept' of a command and the fields (dialogue elements), i.e. the command contents. The fields are **selection**, e.g. display, select; **order**, e.g. altitude, flight-level, heading; **qualifier**, e.g. plus, down right; **numerical**; **action**, e.g. off, arm, climb. By using the contextual information the recognizer could establish what was needed to complete a command if it or the pilot made an omission or a single recognition error occurred. This type of system allows the pilot some flexibility in his/her dialogue.

Williamson *et al.* (1987) concluded that current speech technology was too restrictive and did not allow the pilot enough freedom from the equipment to reach its full potential, and that speech technology needs to understand and respond to the pilot as if it was another pilot. In order to achieve this, the aircraft needs knowledge of the external situation, mission goals and pilot preferences. This kind of intelligent integration between the system and the aircraft is not close in developmental terms. Williamson *et al.* (1987) stated that this type of system will result in a 'pilot-vehicle interface that is flexible, reliable and effective in improving the pilot's overall capabilities in tomorrow's aircraft'.

As the dialogue process is two-way the pilot will require some feedback so that s/he is aware of the status of the recognizer and what it has 'heard'. The major elements in the equation will be the amount, nature and mode of feedback required by the task and the information type involved in the interaction. For example, if speech is being used to control the format on a flight deck display, feedback is generated simply by the display changing, and other feedback is not required. Although paying attention to the screen requires the pilot to be head in, if the pilot has asked for a format change s/he must want to look at the new information anyway, and the visual change will show the command has been received or carried out. If an error occurs in the input or recognition process it will be immediately evident when the wrong format appears, or no change to the format occurs. Some tasks such as data entry may require additional feedback as the changes to the system are not directly evident. It is also the case that some commands have specific forms of feedback already associated with them and it would be inadvisable to modify these without very specific and tested reasons. For example, 'land gear down and locked in position' on a commercial flight deck is always indicated by three green lights.

In the early days of discussions and research into avionic applications for speech recognition it was often argued that voice should also be used as the medium for feedback in order to maintain the eyes-out, head-up operation. However, there are some drawbacks to the use of speech output on the flight deck or in a military cockpit due to the high noise levels. For example, all messages relating to air traffic control are passed by voice, in both military and civil aircraft, and this involves the pilots having to monitor one or more radio channels constantly listening for information for them or relating to their aircraft. Thus, speech output messages used as feedback may cause interruptions or be missed. The number of crew on the flight deck is another compounding factor. Most civil aircraft operate with two or three crew members, whereas many military operations are undertaken by single pilots. In a multiple crew situation voice feedback may be unsuitable due to confusion and masking by human conversation and interaction between the crew members, and the additional pair of eyes allows more freedom for one crew member to spend brief periods head down to receive visual feedback. In military jets the heavy demand for head up eyes-out operation may mean voice feedback may be optimal even though the requirement for heavy monitoring of radio channels during some missions would seem to indicate it was not the best solution. Visual feedback may be particularly acceptable in military jets and increasingly some civil transport aircraft have display areas in line with the 'windscreen' known as a 'head-up display' (HUD) and on which vital flight phase information is presented. It is feasible to use part of this screen as a scratch pad that could be utilized for the presentation of visual head up, eyes forward feedback.

Little research appears to be available in the area to help specify guidelines for the provision of feedback for a speech recognizer, possibly as task element is a strong factor in the decision process for feedback modality and content decisions. Recent research based on a civil flight deck using speech output as feedback to tell the pilot the status of the recognizer syntax at the point of a mode change indicated that trial pilots felt it unnecessary to have spoken feedback for this application (FANSTIC: Synthesis Report). They did not need to be told the system was changing as other visual and system clues were available. It was later when they returned to the system and could not recall the current set up of the recognizer that feedback was required. However this could quickly be gained by visual inspection of the recognizer panel.

In order to obtain and maintain ease of interaction and a reduction in workload, the recognizer dialogue must be as error free as possible as errors will only serve to break up the flow and continuity of the interaction. Error protection and recovery within the dialogue must also be simple, straightforward and easy to mange. The dialogue for the interaction with any system by any input device must allow for the correction of errors made by either the system or user. Enterkin (1991) found that using voice for correction took twice as long as when it was undertaken by keys. The selection of vocabulary for correction is also a cause for concern as it is important that error correction does not inadvertently occur during an input. The words *eight* and *delete*, and *insert* and *enter* were found to cause recognizer confusions when concurrently active during the entry of data strings (Enterkin, 1991).

Performance requirements

The performance of the speech recognition system can be measured in a variety of ways, e.g., by individual word recognition rate, or commands recognized - a correct recognition only being scored if all words were correctly recognized and the input string completed successfully. Add the user to complete the interactive system loop and the measurement will include his/her ability to recall the dialogue content and structure for entering commands, etc. This will also affect performance. The situation is further complicated by the domain used for testing, and whether it is laboratory, simulator or real flight.

A further issue relating to performance measurement is concerned with establishing the rate of performance required for successful application to an avionics task. Humans make many mistakes in speech such as mispronunciation, slips of the tongue and muddling words (Enterkin, 1991), and some may question why should the recognizer be better. Not only do humans make speech errors, their use of keyboards and other input devices are not error free, and interactions involving the use of keyboards show error rates. The question centres around how perfect the speech recognizer performance has to be, given its additional benefits. Here the nature and criticality of the task has to be considered, as the more critical the task, the higher the performance requirement. For example, as the most critical task is to keep the aircraft in the air and on course, maybe performance should be measured in terms of the effects of using the input device on performance of these tasks, with the control of systems being only secondary tasks. This leads to the problem of workload definition and measurement, which it is inappropriate to discuss at length here. Factors needing consideration include: time pressure; task difficulty; other concurrent tasks; overall task objectives; mental/physical effort and combination of effort types; frustration; stress and fatigue. The FANSTIC programme research concluded that speech recognition should not be applied to urgent critical tasks where rapid responses are needed, as it is far quicker to push one button than it is to speak a command. However, for tasks which are not time critical the use of speech may be beneficial and acceptable even if the task is still critical in nature, e.g. lowering the landing gear.

White *et al.* (1984) discussed the multiple performance criteria required. These included: the rate of word substitution; words being incorrectly rejected; levels of noise acceptable by the recognizer as part of the dialogue (a false acceptance); ability to reject non-vocabulary words; the number of words active at any one time.

Much research has been undertaken into setting up databases and procedures for testing recognizers (Taylor, 1987). Many people have stated that speech recognition would be the preferred input modality in the military jet if a consistent recognition rate of 97 per cent and above, for a vocabulary of isolated words and digits was achievable, but this has yet to be proven (Galletti and Abbott, 1989). The major questions, of how good does the recognizer really need to be and how performance is measured still seem to remain unanswered.

Environmental issues

As already stated much of the research has centred on the application of speech technology to military aircraft, and in particular fast jets. This domain has some special features which impact on the use of speech technologies, and these are discussed below.

Much of the work with fast jets and helicopters undertaken in the early 1980s (Beckett, 1982, 1985, 1986; Mountford and North, 1980; Porubcansky, 1985) looked at the effects of the environment, i.e. noise, high 'G' levels, and the use of oxygen masks. Cockpit noise is a problem that is particularly relevant to the military domain, but civil flight decks can also be noisy. Noise in the aviation environment comes from several sources - military aircraft suffer high noise levels from engines, etc., while the crew wear oxygen masks which have their own set of sounds all of which can be 'heard' by the recognizer and have to be dealt with. Noise levels measured in a Bell LongRanger helicopter varied from 90 dBA to 95 dBA over a variety of manoeuvres. Recognizer accuracy tests (Coler, 1984) carried out on a 20-word active vocabulary in this environment resulted in recognition rates between 95.8 and 98.3 per cent depending on the manoeuvre. Noise rates in other aircraft can be even higher, e.g. the Chinook CH-47 helicopter records 115 dBA of noise in cruise, and the F-16 research programmes consider noise levels up to 110 dBA (Taylor, 1987), although peak noise levels can reach up to 140 dBA in some aircraft.

In both the military and civil world the aircraft cockpit is far from a vocally silent environment, even if flown by a single pilot. All instructions relating to the passage of the aircraft through airspace are currently passed to the pilot via voice. Some military and all civil aircraft have two (or more) crew members talking, which is both task oriented and social. All speech on the flight deck will cause difficulties for the recognition process, as the recognizer will try to recognize any incoming sounds. This can lead to errors and in particular, 'addition' errors. If the dialogue is not robust this could change the state of the recognizer without the pilot's knowledge, and in the worst case scenario, an unintentional command may be entered. One of the factors in how much extraneous noise enters the recognizer is the location of the microphone, as a microphone located on the interior of the aircraft will pick up far more extraneous speech than a headset one which is located in front of the pilot's mouth. In military aircraft the crews already have a microphone located inside their oxygen mask for other verbal communications, and while this microphone may not be ideal, it is constantly worn and could be utilized for speech recognition purposes. The oxygen mask carries its own difficulties for the recognizer of breath noise from the gas flowing from the regulator through the valves in the mask. However, creating templates of these noises for the recognizer so it can disregard them seems to be a successful approach to handling this source of noise. Noise cancelling microphones are used by some helicopter forces. These have two parts, one which samples the noise in the cockpit and the other which samples noise and speech. The information from the two sources is then processed so that much of the cockpit noise is cancelled out, thus enhancing the speech signal.

A number of techniques have been used in this context to try to overcome noise problems. For example, providing the recognizer with templates of specific noises which when matched are then disregarded, such as non-speech and the click of an oxygen mask. Another method is to introduce a noise-masking algorithm into the recognition process which uses samples of speechless noise 'heard' in the cockpit, and then modifies the speech and noise signal to clean it up.

Another environmental issue specific to the military domain is the effect of acceleration (G's) on the pilots' ability to speak and the recognizers' ability to recognize. During high G manoeuvres hand movements are difficult to make, hence speech may have a significant advantage. Airborne tests in an F-16 to investigate the effects of G on recognition performance used a vocabulary of 36 commands and templates trained on the ground with the aircraft engines running. The results showed that under constant G, there was a lack of significant degradation in recognition performance up to four G with only a small decrease occurring at five G, and by seven G the performance had severely reduced. However, this means that for most of the flight envelope G will not cause a problem for speech recognizers.

Other environmental factors that may affect recognition on a transitory basis are stress and fatigue (White, 1982). These factors will vary over time but their direct effects on recognition performance in aircraft is not known.

Training

Current recognizers require templates of some kind to which incoming speech is matched. This is one of the most critical elements in the speech recognition process (Barry and Williamson, 1986), because without good templates acceptable recognition performance is not possible. Templates are required for each 'word' as defined by the recognition system. This is normally the same as a spoken word, but in the case of some large vocabulary systems, speech elements such as a syllable are recognized and a dictionary compiling system is then used to select the word spoken.

Most current systems require each user to give sample(s) of the vocabulary items required as templates. The number of passes of each word, whether these are amalgamated to produce an average template or held as individual items, or taken from isolated input or embedded within speech, is dependent on the recognizer involved. However, the problem for airborne application of the technology is that each user requires his/her own set of templates. Military aircraft have an operational requirement for immediate 'take off' so training on the ground prior to a flight would not be feasible. In addition, as the scope of the applications are increased the time required to lay down a set of templates would be prohibitively large and tedious if it had to be repeated each time a pilot flew. Therefore, templates need to be stored for use, but in both the military and civil domain each aircrft is flown by a variety of pilots. Hence, templates for one pilot would have to be stored on all aircraft that s/he may fly and conversely all aircraft must carry the templates of all possible pilots.

Another solution would be to store a pilot's templates on a portable device that s/he takes to the aircraft at the start of a flight. This is technically feasible with a device roughly the size of a credit card, with one possible storage technology for this being electronically erasable programmable read only memory. This means templates can be stored, carried and downloaded as required, and if the templates change or require updating the old templates can be removed from the card and the new ones recorded onto it. A different solution, which may be useful as the size of vocabularies increases, is to introduce an algorithm that has the capability of monitoring incoming speech and modifying a baseline set of templates in the light of previous successful recognitions. The ability for the templates to adapt may be useful in terms of an operation tracking vocal changes such as those due to stress. The use of this type of technique back on the ground may allow the adaptation and development of templates which are more robust for the aircraft environment.

Besides training issues associated with the recognizer technology, the user must also be trained in the use of the recognizer to obtain maximum overall performance. It has been observed that familiarity with the recognizer can affect production of templates (White and Beckett, 1983). Inexperienced individuals have been found to speak differently during training sessions than under operational conditions, hence decreasing the chance of a good match and reducing recognition performance. Therefore, the templates used during operation should be updated once the user has become familiar with the device type, system and interface dialogue.

Conclusion

Research to date has indicated that there are many benefits to be attained from the introduction of speech technology to the avionics environment, particularly in the high workload of the military fast jet. However, the technology still has high cost and risk factors associated with it – only when the advantages are further realized, as flight deck systems and their interfaces become increasingly complex, might widespread introduction of the technology occur. We are still a long way from being able to interact freely with computers by voice (Gorden, 1989). Naturally language would offer a fully flexible interface, which would also be useful for many applications more complex than the tasks carried out in the aircraft cockpit. For the application of current or near term recognition technologies two issues remain. Guidelines for the design of the interface are not yet clearly laid down so producing the optimal interface with speech would require more research, and in the interim period, other technologies such as touch and gesture control are being developed complete with their own set of advantages. The competition for the control of information and systems on future flight decks is still a very open one.

References

Barry, T. P. and Williamson, D. T., 1986, Software methodology for automated recognition training, *IEEE*, 0547-3578/86/0000-0799.

Beckett, P., 1982, A comparison of data entry methods for a single seat aircraft, M.O.D. Contract A7a/569, BAe/FS/R44.

Beckett, P., 1985, Voice control of cockpit systems, Phase 1 Final Report, M.O.D. Contract A85c/3132c, BAe-WFS-TR-RES-221.

Beckett, P., 1986, Voice control of cockpit systems: Feedback to the pilot Phase 2 Final Report, BAe-WFS-TR-RES-294.

Coler, C. R., 1984, In-flight testing of automatic speech recognition systems, *Speech Tech '84*, **1**(1), 95–8.

Enterkin, P., 1991, Voice versus manual techniques for airborne data entry correction, Proceedings of the Ergonomics Society 1991 Annual Conference, *Ergonomicss - Design for performance 1991*, E. J. Lovesey (ed.), London: Taylor & Francis.

FANSTIC Task 4.0 Final Report, 1992, SA-4.0-SR-0001.

FANSTIC Task 4.1 Synthesis Report, 1992, SA-SR-001.

Galletti, I. and Abbott, M., 1989, Development of an advanced airborne speech recognizer for direct voice input, *Speech Technology*, **5**(1).

Gorden, D. F., 1989, Voice recognition and systems activation for aircrew and weapon system interaction (source unknown).

Laycock, J. and Peckham, J. B., 1980, Improving piloting performance whilst using direct voice input, RAE Technical Report 80019.

Mountford, S. J. and North, R. A., 1980, Voice entry for reducing pilot workload, *Proceedings of the 24th annual meeting of the Human Factors Society*, Los Angeles, CA, pp. 185–9.

Porubcansky, C. A., 1985, Speech Technology: present and future applications in the airborne environment, *Aviation, Space and Environmental Medicine*, **56**(2), 138–43.

Starr, A. F., McGuiness, B. and Tyrie, R., 1992, Is voice the right answer for the avionic environment? A comparative study of three input devices. *Proceedings of the Interactive Speech Technology Conference*.

Taylor, M. R., 1986, Voice input applications in aerospace, in *Electronic Speech Recognition, Techniques, Technologies and Applications*, **15**, G. Bristow (ed.), London: Collins.

Taylor, M. R., 1987, Assessment of speech technology in a multi-lingual aerospace environment, *The Proceedings of International Speech Tech '87*, Media Dimensions, Inc.

Warner, N. and Harris, S., 1984, Voice-controlled avionics programs, progress and prognosis, *Speech Tech '84*, **1**(1), 110–23.

White, R. G., 1982, Automatic speech recognition as a cockpit interface, RAE, Technical Memo. FS(F) 454, London: HMSO.

White, R. G. and Beckett, P., 1983, Increased aircraft survivability using direct voice input, Technical Memo. FS(F) 515, RAE, Farnborough, UK.

White, R. W., Parks, D. L. and Smith, W. D., 1984, Potential flight applications for voice recognition and synthesis systems, *Sixth AIAA/IEEE Digital Avionics System Conference*, 84-2661-CP.

Williamson, D. T., Small, R. L. and Feitshans, G. L., 1987, *Speech Technology in the Flight Dynamics Laboratory*, NAECON, 897–900.

10

Listening typewriters in use: some practical studies

I. R. Murray, A. F. Newell, J. L. Arnott and A. Y. Cairns

Introduction

The development of automatic speech recognition machines has been an active research area for the speech community for many years, and one goal of great interest is that of speech-driven word processors, often referred to as 'listening typewriters', for creative writing tasks. As the performance of speech recognition machines is not yet sufficiently high to build a practical system, simulations using a concealed typist operating as a speech recognition machine have been used to assess potential performance.

Gould *et al.* (1983) simulated a listening typewriter in an extensive study using a qwerty keyboard with operators acting as speech recognizers (though only primitive editing facilities were provided), and concluded that 'people will probably be able to compose letters with listening typewriters at least as efficiently as with traditional methods'. Some of their findings have been questioned (Johnson, 1987), and the research project described here extended the Gould simulation. The two main advantages of this project were that the interpretation of the user's commands were performed by a computerized natural language interface (rather than by the secretary), and the user was allowed to use natural speech rates (up to 200 wpm), rather than speech rates constrained by the use of a qwerty keyboard (up to 60 wpm). To permit these natural speech rates, the input system was based on the palantype machine shorthand transcription system developed jointly by the Universities of Southampton and Dundee (Dye *et al.* (1984)) and marked commercially by Possum Controls. With this system a trained operator can produce orthography from verbatim speech with approximately 95 per cent accuracy.

The speech-driven word processor system was used to examine human factors implications of the use of the speech modality, including design of appropriate dialogue structures and the human interface, and produce requirement guidelines for automatic speech recognition machines.

The listening typewriter

The listening typewriter simulation was implemented using an IBM PC-AT palantype-to-orthography system, and a Sun 3/160 workstation which processed the user's dialogue as well as the screen text editor which the user viewed on the Sun monitor (see Fig. 10.1). Unlike Gould's system, which used an 'intelligent' typist to carry out the editing commands, the palantypist in the present system only typed the user's exact words; processing the dialogue and performing the editing instructions was to be performed by the computer. The NLI software developed and used in the present system was implemented in Prolog, and accepted a limited subset of the English language as an input (in ASCII text form from the transcription system). The natural language interface permitted the use of much more complex editing commands than the earlier Gould-type system. For example, the following commands could be understood by the system:

- *'Delete this word'*
- *'Capitalize the first letter of the second paragraph'*
- *'Replace the next two words with the word spelt WEAR'*
- *'Insert an apostrophe after the fourth letter of the third word on the first line of the second last paragraph'*

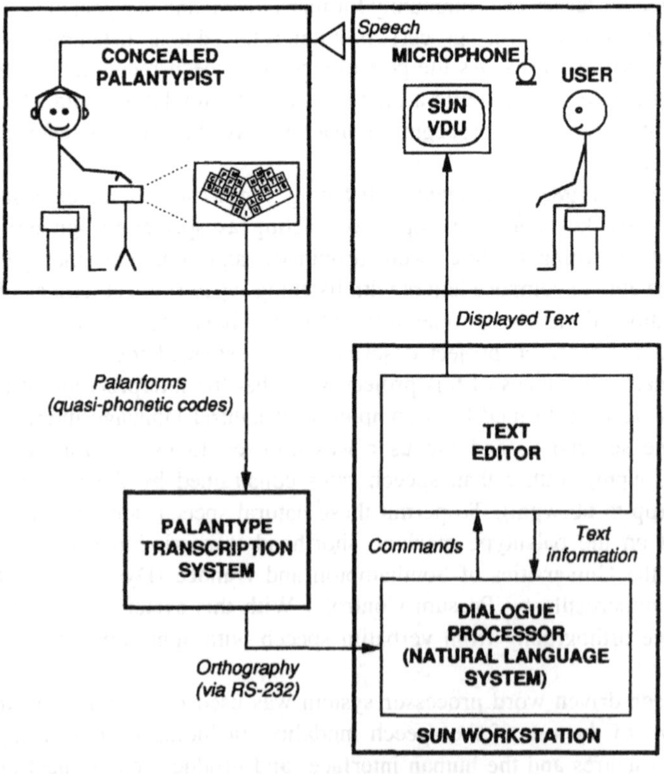

Figure 10.1 The palantype-based 'Wizard of Oz' simulation.

However, combined instructions such as:

- *'Delete this word and replace it with the word NEW'*

were generally not permitted.

Evaluation of the listening typewriter

The 'listening typewriter' system was tested extensively in a series of experiments to evaluate its performance and characteristics, as well as evaluating changes which were made to improve the system. In all of the experiments, the subject's task was composition of a document on the text editor screen. The actual document type and composition mode (copy typing or the subject's own words) varied between experiments. All subjects were asked to complete a questionnaire about the experiment at the conclusion of their session.

Pilot experiments

Two pilot experiments were run to provide practical experience in the experimental procedures required, and an initial understanding of the sorts of spoken commands which a user would employ in this environment. Totally unrestricted dialogue and natural language input were allowed; the operator acted as a natural language processor, listening to the instructions given by the subject and translating them into the keyboard commands of the word processor; the editor was controlled by a standard qwerty keyboard rather than by a palantype keyboard. In these pilot experiments, allowing 'natural language' commands was found to be more of a hindrance than a help, and, despite the typist interpreting the commands and filtering out unwanted input, the users were not very impressed with the system.

Experiments 1 and 2 – partial replications of Gould's simulation experiment using natural speech rates

The first two formal experiments were intended to partially replicate the earlier work by Gould *et al.* (1983). While Gould's experiments were based on a querty keyboard-driven editor, use of the palantype transcription system meant that full-speed entry was possible in this project, although only similar primitive editing commands were available. In order to investigate any effects caused by using a simulation, half of the subjects in this experiment were told about the existence and purpose of the palantypist (overt group), and for the others, it was implied that the system was a fully operational speech recognition machine (covert group).

The composition rates achieved were lower than those achieved by Gould (despite the faster transcription speed); the subjects all rated the system as worse than writing, and were less impressed by the system than Gould's subjects. Most notably, the subjects who knew they were using a simulation rated the system higher and were more impressed than those who were led to believe that they werre talking to a machine. These differences may have implications for further simulation studies, and may be relevant to how 'human' any future automatic speech recognizer would appear.

Subjects tended to dictate only one or two words at a time (despite the potentially high transcription speed), apparently to avoid errors, and it was hypothesized that speech rates would have been higher if no editing had been performed. Thus, in an attempt to increase the composition rate, experiment 2 repeated the covert case of experiment 1, but the subjects were told to ignore any errors made; one document was composed using the previous unrestricted editor, and one using a restricted editor (with formatting commands only). The speech rates were found to be higher than in experiment 1, but higher for the unrestricted editor than for the restricted editor; the hypothesis above was thus not supported.

Experiment 3 – gathering examples of typical user dialogues

The purpose of experiment 3 was to gather examples of subjects' verbal corrections of text as a precursor to developing the natural language pre-processor for the text editor. The subjects were presented with a series of paragraph pairs on the screen, and were asked to speak editor commands to change one paragraph to make it identical to the other; the edits were not performed on the screen. Most subjects found the task of initiating oral commands to an unresponsive machine quite difficult, and most evaluated their own commands as being unclear and at times ambiguous; additionally, the lack of any restriction on commands appeared to be a hindrance to many subjects.

Experiment 4 – longitudinal static experiment

Experiment 4 was an in-depth case study using the listening typewriter including an automatic natural language parser. The five subjects were all retired male executives unfamiliar with word processors (although three were experienced at dictating to a secretary), who performed precis tasks and letter composition tasks. The system used for the experiment had three modes controlled by spoken natural language: text entry mode, command (edit) mode, and symbol (spelling) mode. The latter two modes used an additional window for the composition of the edit command prior to it being executed. By comparing measurable performance and impressions of the system it was hoped that the experiment would show how easy the system was to learn to use, and show the general acceptability of speech as an input modality.

On average 80 per cent of commands were correctly executed in the subjects' first session, but this increased to 94 per cent by the final session. The average composition rate achieved in this experiment was 4.6 wpm, with little difference between the experienced and inexperienced dictators. The subjects indicated that natural language commands were a slow way of editing text, and in particular that cursor movements with the voice were difficult to perform. To offset the slowness, some subjects tended to rush some editing tasks, failing to check edit commands before executing them, hence producing other errors and paradoxically compounding the slowness. The dictators indicated that the system was a little better than writing, although the two subjects who were familiar with using a dictaphone indicated that it was better than a dictaphone, but similar to or worse than dictating to a secretary. Between the initial and final questionnaires, the subjects' rating of the system improved slightly. The three subjects who performed all ten experimental sessions enjoyed the challenge of mastering the speech-driven word processor, and were generally impressed by the ability to see the spoken word almost instantly.

Experiments 5 and 5A – the effect of different feedback strategies

As composition rates in previous experiments had not exceeded 12 wpm, experiment 5 was carried out to compare the effect of various feedback strategies on composition rates. Subjects were asked to compose documents using systems with normal feedback (i.e. words appearing on screen as quickly as possible), feedback on syntactic marker (i.e. text appeared at the next full stop, period etc.), feedback on request (i.e. text appeared when the subject said *Display*) and mixed feedback (combining the latter two modes). It was hypothesized that the psychological effect of seeing speech as it was dictated would hinder the subjects' dictation speed, and consequently that speed would increase if immediate feedback was not given. The screen layout of the editor was improved from that used in experiment 4, as users' performance in that experiment had indicated that some aspects of the interface were not ideal.

Initial results indicated that the feedback strategy used had an effect on the composition rate. It was possible, however, that the extensive editing performed during composition of some of the documents had biased the results, and the experiment was repeated using different document outlines to limit the amount of formatting required, and this experiment (5A) did not confirm the initial results.

In neither experiment did composition rates exceed 10 wpm, and the average speech rate was less than 33 wpm. The efficiency of dictation was generally low, but improved slightly in the second experiment. Overall, the feedback strategies did not greatly affect composition rates. The subjects rated normal feedback as the most favourable, with syntactic marker feedback the least favourable. Most subjects thought normal feedback a hindrance if they were sure of what they wanted to say, but noted that, with the other strategies, it was possible to 'loose the thread' of what they were saying in mid-sentence. A sub-sample of subjects who generated and then edited their documents was selected, and the corresponding timings noted. This revealed that the subjects spent more time dictating than editing their documents, although the subjects thought the opposite to be the case.

Experiment 6 – the effect of different cursor movement modalities

Experiment 6 was carried out to compare the effect on composition rate of various cursor movement modalities (a touch screen, mouse and voice input) as part of the speech-driven word processor. The results showed that the subjects' speech rates were approximately equal in all three conditions, although the composition rate was lower with speech for cursor control than with mouse control, and using the touch screen produced the highest rate. No subject exceeded a composition rate of 10 wpm at any time. The subject's own preference ratings of the systems indicated that the touch screen was the preferred system, and the speech-controlled cursor was least favoured. These results confirmed earlier suggestions that speech is not the best modality for cursor control and similar functions.

Experiment 7 – adding voice input to a conventional text editor

Experiment 7 was a pilot experiment conducted to compare a conventional text editor with one with the additional facility of speech input (for text entry and formatting – no editing was possible by voice). A multi-modality editor was devised, which allowed cursor control performed by touch screen, keyboard, or mouse.

The average composition rate in the no-speech condition was 13.9 wpm, which was slightly faster than the average rate in the with-speech condition, in which the subjects spent approximately equal times using speech and using the keyboard. It appeared, however, that the higher speed of spoken text entry could have been offset by the editing time necessary due to the larger number of errors than would have occurred with keyboard entry.

Experiment 8 – comparison of the speech-driven word processor, keyboard-driven word processors and a dictating machine

Experiment 8 was carried out to compare document composition rates between four systems: the speech-driven word processor (exclusively speech input), a dictation (into a dictating machine) and editing process, a simple keyboard-driven text editor, and the subject's normal word processor.

The results showed that composition using the speech-driven word processor was significantly slower than all of the other systems; it also had the lowest composition efficiency, with around three times as many edit operations performed as occurred with the keyboard-controlled text editor. Most subjects used a 'dictate-then-edit' strategy rather than an 'edit-mistakes-as-you-go-along' strategy, and by timing these separate modes, it was noted that these subjects spent approximately three times longer editing than dictating.

Observations and guidelines

The series of experiments reported here produced a great deal of data and experience of the practical problems of using speech-driven word processors. On the basis of this, some general observations can be made.

General comments

The human interface and dialogue characteristics are a vital part of any speech input system. Inadequate design of either of these will lead to a very inefficient system.

For composition tasks, the recognition accuracy (speech-to-orthography) must be very high; 95 per cent is barely acceptable.

The operators of speech input word processors (even with natural language command structures) will need significant training in order to use speech efficiently in document creation tasks.

Speech-only listening typewriters are slow and inefficient, and are thus likely to only be acceptable in situations where 'hands-free' input is absolutely essential.

Natural language input

Unconstrained natural language is too ambiguous and inefficient to be appropriate for tasks such as text editing.

Making cursor movements and describing locations in the text is very difficult to do with speech alone, and other cursor control modalities should be available if 'hands-free' operation is not essential.

When faced with a natural language understanding system, operators tend to try to develop a subset of commands. Operators tend not to use articles, conjunctions or prepositions, can be sloppy in their usage of tenses, and repeat words unnecessarily.

Both computer naive and experienced computer users find it difficult to invent appropriate spoken commands for editing operators; few subjects even claim to be using natural language structures.

Using a speech-based practical natural language command system can be stressful, particularly in the early stages of learning.

Human-computer interface

Any visual feedback of spoken words tends to slow down the operator's speech, but long term training may reduce this effect.

Careful design of the screen layout must emphasize the mode in which the system is operating, as subjects tend to concentrate on the area of the screen where their words are appearing.

Some subjects tend to make audible spoken asides, which then appear as text on the screen, and have to be deleted, but it is possible that subjects would learn to avoid these with practice.

Subjects' responses

The response to speech input can be very polarized, with some subjects being very positive and some very negative.

Despite the listening typewriter's low performance in terms of speed and efficiency, many subjects enjoyed using the system in the experimental situation.

In the Gould replication, those subjects who knew that they were using a simulation were more impressed by the simulated system than those who thought they were talking to a machine.

Other considerations

All of the experiments in this project were performed with the subject alone in an office. However, a real office environment might be inhibiting to the use of such a system.

The fully speech-driven word processor was significantly slower for creative writing than either a simple keyboard-driven text editor or a dictation process.

As an ASR system is likely to make more errors during text entry than would a keyboard-driven editor, any speed advantage gained by dictating the text could be offset by the increased editing time to correct the larger number of errors.

Conclusion

The simulation of a full-speed listening typewriter has proved to be a very valuable research tool for the investigation of the human factors of speech and natural language input systems. The individual experiences of the subjects during these experiments and their attitudes towards the system have also provided valuable data concerning the acceptability of such systems, and the potential pitfalls in the design of speech-operated and natural language-based systems.

A detailed description of the speech-driven word processor and its evaluation is given in the final project report to the Science and Engineering Research Council (Dundee University Microcomputer Centre, 1990).

Acknowledgements

The work reported in this document was carried out at Dundee University Microcomputer Centre between 1986 and 1990, under the Alvey Directorate project number MMI/SP/079 (SERC numbers GR/D 3009.9 and GR/F 7059.4).

Thanks are expressed the monitoring officer, Dr Roger K. Moore of R.S.R.E., Malvern, and to Mr D. Gemmell and Dr Colin Brooks of Possum Controls Ltd., Middlegreen Trading Estate, Slough, Berkshire, who were the industrial collaborators on the project. Thanks also to Dr John Bridle of R.S.R.E., Malvern, and to all others who have contributed to the project.

References

Dundee University Microcomputer Centre (1990), Machine shorthand as a full-speed speech recognition simulation, Final project report to the Alvey Directorate (summary in University of Dundee Computer Science Report CS 91/09); a 15-minute video presentation describing the project and showing the listening typewriter in operation was produced to accompany the report.

Dye, R., Newell, A. F. and Arnott, J. L., 1984, An adaptive editor for shorthand transcription systems, *Proceedings of the 1st IFIP Conference on Human–computer interaction* (North Holland), Imperial College, London, **2**, 92–6.

Gould, J. D., Conti, J. and Hovanyecz, T., 1983, Composing letters with a simulated listening typewriter, *Communications of the ACM*, **26**, 295–308.

Johnson, C., 1987, Speech technology and word processing, *Proceedings of Speech Tech '87* (Media Dimensions Inc., New York), London.

Further details on the experiments reported here can be found in the following publications;

Carter, K. E. P., Newell, A. F. and Arnott, J L., 1988, Studies using a simulated listening typewriter, *Proceedings of Speech '88, the 7th FASE Symposium*, Institute of Acoustics, Edinburgh, 1289–96.

Carter, K. E. P., Cookson, S., Newell, A. F., Arnott, J. L. and Dye, R., 1989, The effect of feedback on composition rate using a simulated listening typewriter, *Proceedings of the European Conference on Speech Technology*, CEP Consultants Ltd., Paris, **1**, 402–4.

Dye, R. and Cruickshank, G., 1988, A system for composing and editing text using natural spoken language, *Proceedings of Speech '88, the 7th FASE Symposium*, Institute of Acoustics, Edinburgh, 1321–8.

Dye, R., Arnott, J. L., Newell, A. F., Carter, K. E. P. and Cruickshank, G., 1989, Assessing the potential of future automatic speech recognition technology in text composition applications, *Proceedings of Simulation in the Development of User Interfaces Conference*, Brighton, 216–25.

Newell, A. F., 1987, Speech in Office Systems, *Proceedings of Speech Tech '87*, (Media Dimensions Inc., NY), London, 47–9.

Newell, A. F., 1987, Speech simulation studies: performance and dialogue specification, *Proceedings of Unicom Seminar, Recent Developments and Applications of Natural Language Understanding*, London.

Newell, A. F., 1989, Speech simulation studies: performance and dialogue in Peckham, J. (ed.), *Recent Developments and Applications of Natural Language Processing*, Kogan Page, 141–57.

Newell, A. F., 1992, Whither Speech Systems, in Ainsworth, W. A. (ed.), *Advances in Speech, Hearing and language Processing*, **2**, J.A.I. Press, 253–79.

Newell, A. F., Arnott, J. L., Carter, K. E. P. and Cruickshank, G., 1990, Listening typewriter simulation studies, *International Journal of Man–machine Studies*, **33**, 1–19.

Newell, A. F., Arnott, J. L. and Dye, R., 1987, A full-speed simulation of speech recognition machines, *Proceedings of the European Conference on Speech Technology*, CEP Consultants Ltd., Edinburgh, **2**, 410–3.

Newell, A. F., Arnott, J. L., Dye, R. and Cairns, A. Y., 1991, A full-speed listening typewriter simulation, *International Journal of Man–machine Studies*, **35**(2), 119–31.

11

Voice as a medium for document annotation

Philip Tucker and Dylan Jones

Abstract

There is a reluctance amongst those undertaking certain types of reading task to change from using the 'hard copy' medium of document presentation to using visual display units (VDUs). This is thought to be associated in part with the difficulties of annotating a document presented in the electronic medium (Wright, 1987). A review is presented of research comparing computer-based annotation media (typed or spoken input) with written annotation of documents. While it is concluded that voice offers little promise of aiding the electronic medium in supplanting paper as the most popular medium for annotation, it is noted that voice has a role in supplementing transcription as a medium for annotation. The usability of voice annotation interfaces is discussed in terms of task specification, interface design, user practice and social cues.

Introduction

The move from paper to electronic media for information transmission affords considerable advantages to the user in terms of time, resources and effort. Yet despite the rise of the concept of the paperless office, there are particular reading activities in which users are still reluctant to make the move from paper to the electronic display of documents (Wright, 1987). Two such activities are proof-reading and making extended annotations, such as when refereeing an academic paper. Wright and Lickorish (1984a) propose that the decision to carry out these activities will be at least partly based upon the ease and efficacy with which annotations of the document can be made. The predominance of paper as the firmly established traditional presentation medium for documents has fostered the equally entrenched tradition of handwritten annotation, usually on the document itself. This review will focus on the role of alternative annotation media in the translation from paper to electronic media and, in particular, the potential role for voice as a medium for annotation.

When a document is presented on a VDU, it takes longer to proof-read than when it is presented in hard copy form, and this has been ascribed to the difficulty of annotating the document (Wright and Lickorish, 1984b). Similarly, in a refereeing task, subjects who are asked to read a document and make comments upon its content, in the form of extended annotations, tend to perform more slowly when the text is presented on a VDU, than when the document is presented in hard copy form (Wright and Lickorish, 1984a). Our own research has demonstrated that writing on the document is preferred as a method of annotation to typing, at least by non-skilled typists. This preference appears to reflect the relative speed and ease with which written annotations were made. Reading tasks that involve writing annotations on the document are performed quicker than when the annotations are typed into a keyboard. Furthermore, writing elicits more annotations than typing (Tucker and Jones, submitted).

A quick and easy method of recording annotations may assist the electronic medium of document presentation in becoming an acceptable substitute for paper. Research suggests that voice annotation can be faster than typed annotation, and may therefore be one such method (Van Nes, 1988).

Annotation production

Wright (1987) suggests that voice would afford greater advantage to the making of extended annotations of the type made when refereeing, than it would to proof-reading. This suggestion is based upon a theory of proof-reading strategy proposed by Wright and Lickorish (1984b), that the proof-reader is often able to continue reading the text while brief marginal annotations are being made. On this basis, a voice input to a computer-based editing system would be likely to cause greater interference to continued reading of text than would the recording of a written symbol.

In free annotation tasks that involve making extended annotations, comparisons of voice and typing as means of recording annotations suggest that users make approximately equal number of annotations in either mode (Van Nes, 1988; Tucker and Jones, submitted). Voice annotations contain many more words than the typed annotations (as much as double), but the voice annotations take significantly less time to make. Van Nes states that approximately equal amounts of information are conveyed through each medium, although this presumably refers to linguistic information and does not include para-linguistic information (e.g. stress on particular words).

If, as we have proposed, a quick and easy method of annotation is preferred, then the speed advantage associated with voice annotation should be reflected in the ratings of user preference. Indeed it is the case that in the Van Nes study, eight out of twelve subjects preferred voice recording to typing annotations. However, our own study failed to replicate such a clear preference. Although the method of document presentation differed between the two experiments, it is unlikely that this would explain why voice was preferred in the former study while typing was preferred in ours. Rather it would predict, if anything, the opposite effect. In the Van Nes study, the typed annotations were made upon the same screen that displayed the original document; in our experiment typed annotations were made, on a VDU, of the document which was presented on paper. This difference would have contributed to making the typing task easier in the former study than it was in the latter, because a common focal point for reading and annotating was retained in the Van Nes study, while in our study the two focal points were separate. Wright and Lickorish (1984b) provide further discussion of the effects location of annotation interfaces, relative to the original document.

Any explanation of an effect of annotation mode must consider the exact nature of the reading task involved and the type of annotation associated with it; the choice of optimum annotation mode is likely to be determined by the type of annotation that a particular task involves. It might be suggested that the type of annotations elicited in the Van Nes study were more suited to voice annotation and less suited to typing than was the case in our study. This proposal is weakened however, when it is noted that, unlike our experiment, the task used in the former study included an element of proof-reading and, as previously noted, Wright (1987) suggests that voice is likely to be less suited to proof-reading annotation than for making extended comments.

The document used in the Van Nes study was considerably shorter than that used in our study, and thus it is possible that length of document interacts with annotation media to produce variations in user preferences for a particular medium. We can only speculate as to the differences between the subject samples of the two experiments. It is possible, though unlikely, that the subjects in the Van Nes study were even less skilled in typing than the subjects in our own study. More likely is the possibility that the subjects in the Van Nes study, in common with those in our study, had little previous experience of voice annotation, prior to the experiment. However, it is conceivable that they were allowed longer practice sessions, in order to familiarize themselves with the apparatus, before embarking on the actual experimental trials. As a result, they may have been able to develop a greater appreciation of voice annotation. In order to test this hypothesis, we repeated the comparison of the two annotation media in a second experiment, giving subjects longer and more standardized practice trials in order to acquaint themselves with the annotation interfaces. This change in experimental procedure appeared to have a greater effect upon voice annotation performance (in terms of the mean number of annotations made and preference for voice annotation) than it did upon typed annotation. This was taken as limited support for a theory that small amounts of pre-trial practice will improve the utility of voice annotation but not typed annotation.

Reasons cited by users for a preference for voice included that it is faster; it facilitates the use of complete sentences (as opposed to writing them in their short-form) and, less reading is required (i.e. the checking of annotations). It also facilitates the addition of comments without cluttering the page. It has been noted that because voice annotation imposes a relatively low load on working memory it is possible to retain ideas until a whole paragraph has been read before recording them, while it is felt necessary to transcribe annotations instantly. The ease and speed of voice recording is considered by users to encourage greater elaboration (Gould, 1982; Nicholson, 1985; Van Nes, 1988; Tucker and Jones, submitted).

Yet despite the advantages that voice appears to offer over typed input of annotations, the provision of voice annotation facilities may prove to be inadequate to persuade referees to carry out their tasks 'on-line'. While voice appears to share at least some of the advantages of written annotation, they may not necessarily be sufficient to overcome the other disadvantages inherent in contemporary electronic document presentation. Research suggests that users take approximately equal time to read and annotate a document using voice annotation as they take when writing annotations on a hard copy of the document (Gould, 1982; Tucker and Jones, submitted). However, our research suggests that more annotations are made when the user is able to make written annotations on the document, than when voice recording annotations. Moreover, voice was associated with its own set of particular disadvantages for recording annotations. Users generally expressed a preference for making written comments on the document, as opposed to spoken ones, as they were easier to review and were more easily attached to the appropriate place in the text. A degree of cognitive impairment was reported by many users of voice annotation, as a result of a perceived pressure when voice recording comments. Evaluation of the argument (i.e. considering what to say) was felt to be interfering with the effort of thinking how to say it in a continuously spoken sentence. Self consciousness was also mentioned as a factor in some of the negative evaluations of speech. Further manifestations of the pressure that users feel when voice recording included difficulties in closing, which lead to 'rambling'. In a similar vein some users considered being able to record their thoughts rapidly in the voice condition an advantage, while others considered this a disadvantage due to the lack of an intervening process to allow for clarification of thoughts. Users also reported making inappropriate or inconsistent comments and choosing inappropriate words as a result of the perceived pressure.

The pressure may partly be attributable to the absence of any response to the users' utterances by the machine. Although voice annotation messages are a form of human - human communication, the actual act of voice recording is often perceived as talking to the machine. Part of the difficulty associated with voice recording relates to the absence of feedback from the machine of the sort encountered when speaking to another human being. This is a form of feedback known as encouraging phrases. Research has demonstrated the utility of encouraging phrases in enhancing human - computer spoken dialogues (Miles *et al.*, 1989). The provision of intelligent feedback in the form of encouraging phrases generated by the voice annotation device is currently technologically unfeasible. It would however be possible for the device to generate a standard encouraging phrase every time it detects a pause in the speakers input of a pre-specified duration.

Research attempts at qualitatively assessing extended annotations have borne little fruit. In their study of refereeing articles for an academic journal, Wright and Lickorish (1984a) found that comments made were highly variable, as were the levels of evaluation and suggested corrections. As a result of this variability, it is unsurprising that there were no discernible effects of medium on subjects evaluation of content. It was concluded that '. . . given the variability observed, there seems little incentive for collecting more data of this kind.' (Wright, 1987 p. 38). Our own study sought to circumvent this problem by measuring one aspect of what might influence the quality of a referees evaluation; namely their ability to evaluate the authors' argument. This in turn could be determined by testing the referees comprehension of the text in a series of questions given after the referee has evaluated it. In the event, we failed to find any effect of annotation medium upon the users comprehension of the document. From this it was possible to conclude that any decrement in the quality of annotations, such as there may have been, would have been attributable to the expression of evaluative comments, rather than their formulation.

Annotation reception

Two experiments on the reception of annotations in voice and text mode are reported by Van Nes (1988). In one experiment, professional secretaries carried out editing operations upon a document in accordance with instructions contained within annotations in either spoken or text form. Editing performance was degraded most noticeably in the voice annotation condition, if the instruction concerned higher level editing operations (i.e. changing sentences or text structure, as opposed to altering individual words). The effect of annotation mode on performance was reflected in the preference ratings, with preference for text over voice becoming more prevalent as the editing instructions referred to higher levels of text structure.

In the other experiment users processed complex, partly conflicting annotations from four originators. Subjects processed annotations in either mode with equal speed. It is interesting to note that the perceived authority of the annotator had a strong influence on the confidence with which the task was performed; however some subjects objected to such 'bossy' authoritative tones. The receivers of these annotations generally preferred text annotations. Taking all the results from both reception experiments, approximately 75 per cent of subjects preferred text annotation.

Field studies of voice annotation

In a study of the implementation of an experimental voice store-and-forward system, Nicholson (1985) noted that voice was most commonly used for forwarding informal annotations to documents, as well as being used to provide introductory or handling instructions for typed documents (e.g. 'John, please review this and report back on Friday'). Content of the voice messages was short (between 21 and 38 seconds), but this was attributed partly to the system's limited disk storage space. There was no correlation between voice usage and job status. Users' comments detailed a number of advantages that voice afforded to the performance of annotation tasks. In addition to the benefits cited by experimental subjects (see above), users noted that voice adds 'a personal touch', and it enables the conveyance of nuance. Some comments reflected the way in which the introduction of new technology, such as voice annotation facilities, can actually alter aspects of office routine. For example, it was reported that voice annotation was replacing face-to-face explanations and meetings.

Amongst infrequent users of the system, typing was thought to be quicker and more natural. They mentioned the effort of breaking old habits in order to use voice. Concern was expressed about storage requirements by engineer users. There was resistance to voice along the lines of not liking the sound of one's own voice and feeling uneasy on the telephone. Voice messages were considered annoying or intrusive; sending text messages was considered more polite. Users who preferred handwriting to voice expressed sentiments concerning their unfamiliarity with the voice system. They spoke of their ability to think more clearly with written text. It was also noted by users that recipients would have the advantage of having notes to refer to.

Nicholson concluded that the voice store-and-forward system was best suited for short, personal, informal communications, to a limited number of addressees. It was also noted that in annotation, voice is a mode for supplementary information; not a replacement for traditional text information (e.g. formal communication, reference tables, etc.). It was suggested that resistance to usage by occasional users is a result of unfamiliarity, leading to discomfort (to be contrasted with the keyboard's unpopularity amongst non-typists) and that this could be overcome by training.

Conclusions on voice annotation

The possible advantages to the proof-reader, offered by voice annotation over other media, remain largely unspecified. However, unless these advantages are great, and this seems unlikely (Wright, 1987,), voice offers little promise of aiding the electronic medium in supplanting paper as the most popular medium for proof-reading. The recipient of a proof-read document appears, in general, to prefer textual annotation to voice annotation as voice offers them no speed advantage over textual annotation. Furthermore, voice is inappropriate for annotations concerning the recomposition of text at the higher levels of composition (i.e. sentence and paragraph level) (Van Nes, 1988). Voice is not an efficient medium for the description of spatial information (Cowley, Miles and Jones, 1990) and thus may be inappropriate for conveying 'cut and paste' editing instructions. Voice annotations may be more appropriate for short, simple, global comments about the document, such as handling instructions (Nicholson, 1985), as these do not refer to particular parts of the text. The brevity of the voice annotation will benefit the receipt of the annotation, and will make composition easier as memory burden is reduced.

When making extended annotations of the kind, say, associated with refereeing an academic paper, voice would appear to be an acceptable substitute for typed input of annotations, provided that subjects have adequate opportunity to become familiar with the medium. However, it is yet to be demonstrated that voice can serve as an acceptable substitute for writing annotations (Gould, 1982; Van Nes, 1988; Tucker and Jones, submitted). In terms of processing time, the reception of extended annotations appears to be unaffected by annotation modality (Van Nes, 1988) although users tended to prefer textual annotations. Once more it may be necessary to consider the nature of the particular type of information conveyed, before generalizing from this finding.

When a series of communications take place through the annotations of a document or documents, the medium of annotation will affect the nature of these social interactions. Relative to textual communication, voice will convey more cues to the receiver regarding the sender's status and position and thus will tend to reduce social anonymity. Hence inexperienced users of voice annotation feel self-conscious expressing themselves in this way. At the same time voice increases social influence and thus social exchanges by voice messaging may become more person orientated, and less message orientated. It also results in discourse being more regulated and less free. The personalizing effects of voice annotation are potentially advantageous to 'good' decision making, under certain conditions. Rather than aiming for objectivity, this may require the formation of affective bonds, a status distribution that helps sort out multiple objectives, and a hierarchy that determines influence (Kelly and Thibut, 1978; March and Olsen, 1976; Salancik, 1977). Whether the effect of personalization is either negative or positive may depend upon existing relationships between individuals: either exacerbating misunderstandings, or making friendly co-operative exchange more efficient (Kiesler, Siegal and McGuire, 1984).

Voice may have a number of advantages over text for the presentation of information that:

- is of low complexity;
- does not require attention to every detail of the message;
- is intended to set a mood, and;
- is to be high in persuasive power.

Software packages and other 'documents' are often supplied with textual introductory preambles or interpolated commentaries. Rather than go to the effort of reading this material which users often consider superfluous, they will tend to skim or ignore such information. However,the acquisition of this information may actually be crucial to the overall cogency of the document. The reception of voice messages tends to be a quicker, easier, more passive process than reading text, providing the message is simple and well structured. Thus, in an electronic multimedia document, introductions and certain types of commentary may be more effectively presented in speech than in text.

In conclusion, voice should be used as supplementary medium to transcription for annotation. The preceding discussion illustrates the fact that voice is only appropriate for certain forms of annotation. The majority of reading tasks that require annotation facilities will require both forms of annotation media.

Acknowledgement

Philip Tucker is funded by a studentship provided by the Digital Equipment Corporation under the European External Research Programme. Thanks are due to Brian Rees, Jon Barrett, John Brooke and Chris Bartlett of DEC for their continued contributions of ideas and equipment support.

References

Cowley, C. K., Miles, C. and Jones, D. M., 1990, The incorporation of synthetic speech into the human–computer interface, *Contemporary Ergonomics 1990*, London: Taylor & Francis.

Gould, J. D., 1982, Writing and speaking letters and messages, *International Journal of Man-machine Studies*, **16**, 147–71.

Kelly, H. H. and Thibut, J. W., 1978, *Interpersonal Relations*, NY: Wiley.

Kiesler, S., Siegal, J. and McGuire, T. W., 1984, Social psychological aspects of computer-mediated communication, *American Psychology*, **39**(10), 1123–34.

March, J. G. and Olsen, J. P., 1976, *Ambiguity and Choice in Organizations*, Bergen, Norway: Universitetsforiaget.

Miles, C., Jones, D. M. and Simpson, A., 1989, Human factors in speech synthesis: factors affecting friendliness and efficiency, *Proceedings of the Auropean Conference on Speech Communication Technology*, Paris Sept 89, Edinburgh: CEP.

Nicholson, R. T., 1985, *ACM Transactions on Office Information Systems*, **3**(3), 307–14.

Salancik, G. R., 1977, Commitment and the control of organizational behaviour and belief, in Staw, B. M. and Salancik, G. R. (eds), *New Directions in Organizational Behaviour*, Chicago: St Clair Press.

Tucker, P. T. and Jones, D. M. (submitted), Document annotation: to write, type or speak?

Van Nes, F. L., 1988, Multimedia workstations for the office, *IPO (Eindhoven) Annual Progress Report 23*.

Wright, P., 1987, Reading and writing for electronic journals, in Britton, B. K. and Glynn, S. M., *Executive Control Processes in Reading*, Hillsdale, NJ: Erlbaum.

Wright, P. and Lickorish, A., 1984a, Investigating referees requirements in an electronic medium, *Visible Language*, **13**(2), 186–205.

Wright, P. and Lickorish, A., 1984b, Ease of annotation in proof-reading tasks, *Behaviour and Information Technology*, **3**, 185–94.

Part III

12

Considering feedback and error correction

C. Baber

Introduction

It has long been recognized that automatic speech recognition devices (ASR) are prone to problems in consistently recognizing human speech. The underlying causes of these problems appear to stem from the technology itself (although such problems are underway to being eradicated) and from the user of the device. People who use ASR devices often have difficulty in maintaining a consistent speech style. ASR devices tend to work on principles of matching incoming speech with stored samples; any variation between incoming and stored speech will lead to recognition errors. The prevalence of recognition errors are viewed by many as a major obstacle to the use of ASR. The papers in this section demonstrate that recognition errors can be dealt with effectively by a range of error correction strategies.

In order for users to know that an error has occurred, it is necessary to have some means of informing them of the device's recognition performance, i.e., what the ASR device has recognized in response to incoming speech. The issue of providing feedback to users from ASR devices has been covered in depth in the literature (McCauley, 1984; Hapeshi and Jones, 1989). However, in Chapter 13, Frankish and Noyes offer a novel perspective by considering 'user feedback' to the ASR device.

One of the referees of this chapter pointed out that Frankish and Noyes appeared to be using 'feedback' in a somewhat ambiguous fashion; surely, the referee argued, feedback must be unidirectional, i.e., it could only come from the ASR device. This argument would be true if the user and ASR device assumed fixed roles in an interaction. If, however, we view the interaction as progressing through a switching of speaker/auditor roles, then it is possible to view feedback from the user as a meaningful concept. In this context, user feedback would serve as a means of confirming recognition performance, in a similar manner to back-channelling in human–human communication, but would not be a form of command input. In this sense, the information from the ASR device will elicit feedback from the user in relation to its correctness, with incorrect feedback eliciting a different form of feedback. I suppose in this sense even no response from the user could function as feedback in that it indicated an acceptance of the device's recognition performance.

There will be instances, however, in which the user may need to explicitly take action to deal with recognition errors. By convention we tend to distinguish three principal types of recognition error: false insertion, false rejection and substitution (see Foreword). Substitution error represents the most difficult class of problem, and Chapters 14 and 15 consider different strategies by which such errors can be dealt with.

In Chapter 14, Ainsworth and Pratt consider strategies of either allowing an ASR device to suggest an alternative word to the one which was misrecognized, or allowing the user to repeat the word. In the former condition the ASR device produced a second choice word for consideration by the user. The second choice word could share sufficient characteristics with the spoken word for confusion to occur. In the latter condition the ASR device removed the erroneous word from its candidate set, and asked the user to speak again. Removing the word from the candidate set would eliminate a possible source of confusion. This showed that asking the user to repeat the word led to superior performance, especially under noisy conditions. This 'user-centred' approach sits neatly with the Chapter 13 (Frankish and Noyes), and suggests that the user should be in charge of error correction.

In Chapter 15, Murray *et al.* produce a different conclusion. They used strategies which were identical to those in the Ainsworth and Pratt study, and included a further condition of serial elimination of templates. In this latter condition, erroneous words were eliminated from the candidate until the correct word was recognized. Murray *et al.* found that the second choice approach produced better performance.

Thus, there is an apparent contradiction in the conclusions of these chapters. However, it must be borne in mind that the studies employed different ASR devices, different vocabularies, and different experimental conditions. This would suggest that error correction for ASR is not a simple matter of choosing a strategy and applying it. Rather one needs to take into account the range of factors which will affect the interaction between user and device.

References

Hapeshi, K. and Jones, D. M., 1989, The ergonomics of automatic speech recognition interfaces, in D. J. Oborne (ed.), *International Review of Ergonomics*, **2,** 251–290.

McCauley, M. E., 1984, Human factors in voice technology, in F. A. Muckler (ed.), *Human Factors Review 1984*, Santa Monica, CA: Human Factors Society.

13

Feedback in automatic speech recognition: who is saying what and to whom?

Clive Frankish and Jan Noyes

Abstract

In the context of automatic speech recognition, feedback is normally thought of as a process which allows the user to confirm that the device has correctly recognized an utterance; i.e. communication of information from the system to the user. In an 'unintelligent' system, feedback from the user is simply used as a control signal, to initiate entry to error-correction routines. However, the inclusion of error-correction procedures creates a potential for two-way communication of information. The fact that users respond in different ways to recognizer responses that are correct or incorrect allows the system to monitor the accuracy of its recognition attempts. In the short term, this information might be used to facilitate re-entry following misrecognition. Over the longer term, an intelligent system might use the feedback it receives from the user to shape its own behaviour, by means of adaptive algorithms that modify the recognition process. In this paper it is argued that efficient use of the information contained in the transactions between user and system can make a significant contribution to overall system efficiency.

Introduction

In the period since automatic speech recognition became a practical possibility, the issue of 'feedback' has been a major preoccupation of researchers and system designers. The initial reason for this concern is the fallibility of speech recognition devices, which at present are unlikely to achieve recognition accuracies of greater than 95 - 97 per cent under operational conditions. Unless the target application is unusually error-tolerant, some means must therefore be found to signal to the user of a speech input device that an utterance has been correctly recognized.

Few would therefore disagree with the view that 'immediate feedback of recognition results must be given to the user of a voice input system, either visually, aurally or both' (Martin and Welch, 1980). Subsequent research has focused on techniques of providing this feedback that are maximally effective, while interfering as little as possible with the user's primary task (e.g. Schurick *et al.*, 1985; Simpson *et al.*, 1985).

In this research, 'feedback' has generally been thought of as a means of conveying information to the user about identification of an utterance. However, the inclusion of error correction procedures turns this process into a dialogue. In data entry systems that provide concurrent feedback, this two-way transaction occurs for every input. Assuming that the feedback is acted upon, users will respond differentially to correct and incorrect recognitions. This response is itself a form of feedback, conveying information from the user to the system of which the recognizer is a part. Failure to utilize this information may result in a system that is not completely optimized, or one that seriously frustrates the user. In the following sections we consider the implications of feedback from users to the system both for the progress of the current transaction, and also for the longer-term operation of the speech recognizer. This discussion is based in part on data obtained from experimental studies of speech input, using a task that is representative of a particular type of speech input application. While some of the conclusions that follow are restricted to applications of this general nature, others are more widely applicable to the design of speech interfaces.

An experimental task

Although there are many instances of successful application of automatic speech recognition, these tend to involve a rather limited range of tasks and settings (Noyes and Frankish, 1987). In many applications, such as industrial inspection tasks, there is a requirement to capture data at source, under conditions that make keyboard operation difficult or inappropriate.

Many of these tasks involve data entry by trained operators, using a restricted vocabulary. In experimental studies of users' interaction with speech recognition devices, we have modelled these characteristics in simple data entry tasks (Frankish and Noyes, 1990). The data discussed in the present chapter are all drawn from studies using this type of task. In this case subjects used a speech recognition device (a Votan VPC 2000) to enter a series of six-digit data strings. Simple dialogues for data entry and error-correction were included, to enable subjects to correct recognition errors.

The data discussed here were all taken from versions of the task in which visual or spoken feedback was delivered after each user utterance ('concurrent feedback'). In the visual feedback condition, recognition of a digit resulted in the addition of that digit to a screen feedback display. No explicit confirmation was required for correctly recognized inputs. The phrase *try again* could be used to delete the previous digit from the display. After six digits had been recognized, explicit feedback was required to signal completion of the trial.

The procedure for spoken feedback was similar, except for actions relating to the visual display. Feedback was provided by a series of digitized utterances, corresponding to the words and phrases in the recognition vocabulary. Each recognition response was thus acknowledged by a spoken 'echo', consisting of the word or phrase that had been identified.

The role of feedback from the user in error-correction dialogues

This data entry task involves a series of transactions in which the user's input is immediately echoed by the system in some form of feedback display. User acceptance of that transaction can be taken as an unambiguous indication of the user's intentions. In the event of a misrecognition, the task dialogue allows users to reject the entry and re-enter the data. How should the system utilize the information that is now available in order to facilitate re-entry? At this stage the user's intention is still unknown, but it is clear what the user does not want. Failure to build this information into the error-correction dialogue can lead to sequences of repeated misrecognitions that make the resulting system extremely frustrating to use.

We can illustrate this point by considering some data from the experimental data entry task, in which the error-correction was 'unintelligent'. Table 13.1 shows the pattern of recognizer output for first attempts at entry of an individual word, and also the output for the first attempt at re-entry. The pattern of results is very striking. If a word is misrecognized at the first attempt, most second attempts will be successful. However, of those that fail, most will result in an identical misrecognition; this happened for more than 75 per cent of re-entry failures. In other words, users often find themselves being repeatedly offered a recognition response that they have already rejected; a situation that is guaranteed to produce user dissatisfaction and frustration.

Another characteristic of the data shown in Table 13.1 suggests a simple design option that would eliminate this problem. The VPC 2000 identifies the two closest template matches. In most cases, if the first choice is incorrect, then the second choice is correct. If the first attempt has failed, and the re-entry results in an identical misrecognition, it is even more likely that the second choice is correct; this was the case for 85 per cent of the re-entry attempts that fell into this category. The obvious course is to select the second choice template, if the first choice has already been rejected by the user. If this strategy were adopted, the likelihood of a correct recognition on re-entry would increase by 10.4 per cent from 83.8 per cent to 94.2

Table 13.1 First and second attempts at recognition

		Attempt	
Outcome		First	Second
Correct recognition (%)		95.8	83.8
Incorrect recognition; % of cases where 1st choice is the same as for attempt 1:			12.2
2nd choice:	Correct (%)	10.4	
	Incorrect (%)	1.8	
	(% of incorrect responses with 2nd choice correct)		(85)

per cent. This is a significant improvement in system performance, and possibly a crucial improvement in terms of users' perception and acceptance of the system.

More complex options for the utilization of feedback from the user will involve temporary disabling of templates for rejected items. These options are reviewed in Chapter 15. For the present, it should be noted that any strategy of this type must take into account the possibility that the feedback from the user may be unreliable, either because of user error, or because of system misrecognition of the command that triggers the error correction routine. This point will be considered later.

Feedback and adaptive recognition

If errors are always corrected by the user, the device can accumulate information about the recognition process. Once a confirmation response is received, this information can be classified according to which word was actually spoken. The device can thus build up a data base containing information about the pattern of recognition performance obtained with a particular user.

How might this information be used? Two approaches are identified in the following discussion. The first is adaptive template modification, sometimes seen as a strategy for eliminating poorly-performing templates from the matching process. A second possibility is that statistical information about the recent history of recognition attempts might be directly utilized in the decision-making process.

Adaptive systems based on template modification

If multiple templates are maintained for a word in the task vocabulary, it is possible to record how often each template appears as a first-choice match to the corresponding input utterance. After a period of monitoring, the template that produces the fewest successful matches can then be replaced by a template constructed from the next input that the user confirms as an instance of this vocabulary item.

There are two problems that might be tackled by using adaptive systems based on this type of strategy. One is 'voice drift'; changes in speech patterns for a particular user over an extended period of ASR use. An adaptive system should have the capacity to track these changes over time, thus maintaining levels of recognition accuracy. A second application is speaker-independent recognition. A device can be constructed which begins with a set of templates for each word in the target vocabulary, based on speech samples from a representative sample of the population. These templates would be used initially by a new user, but would subsequently be replaced by new templates based on utterances whose identity had been confirmed by the user (Nusbaum and Pisoni, 1986; McInnes *et al.*, 1987). This approach has the advantage that new users could begin to use the system without the need for an initial session of template training.

Although the notion of adaptive systems is attractive, there are attendant dangers. Some method must be found to ensure that templates are not inadvertently updated with data from an inappropriate utterance. With supervised adaptation, this is achieved by careful feedback monitoring and intervention by the user. Unsupervised adaptation requires internal error-checking software to eliminate dubious utterances from the updating routine. The adoption of strict acceptance criteria in the latter procedure can mean that effective adaptation requires high throughputs of all vocabulary items. Another drawback is that confidence estimates for recognition accuracy are necessarily based on matches to the current template set. Although the aim is to improve recognition for speech that has drifted away from the original training tokens, the utterances selected as new templates are likely to be those closest to the original training set. Conversely, if speech patterns and templates do become misaligned, for whatever reason, a conservative updating strategy can delay recovery.

At this stage, the value of adaptive strategies for template-matching devices is somewhat unclear. Several studies have demonstrated enhanced recognition performance in adaptive, as compared with non-adaptive systems (Green *et al.*, 1983; Damper and McDonald, 1984; Graffunder *et al.*, 1985). On the other hand, McInnes *et al.* (1987) found that although supervised adaptation generally resulted in better recognition, performance sometimes declined during trials with unsupervised adaptation.

In a large scale trial reported by Talbot (1987), recognition with unsupervised template adaptation was no better than for a non-adaptive control. The procedure for updating templates in this study was based on the supervised regime proposed by Green *et al.* (1983), but with multiple templates in both adaptive and non-adaptive systems. In their evaluation, Green *et al.* (1983) used several templates for each vocabulary item in the adaptive system, but single templates in the non-adaptive control. This factor alone may have accounted for the observed advantage for the adaptive system.

Adaptive decision rules

The approaches discussed above are based on evaluation of individual template performance, followed by replacement of those that are performing least well. Information derived from monitoring of recognizer performance can also be used in a rather different way, to ensure that the best use is made of templates in the current set. The following discussion is intended to illustrate in principle how this might be done; more sophisticated procedures might well produce a greater payoff in terms of enhanced recognition accuracy.

Template matching and decision making

In common with many recognizers that use template-matching algorithms, the VPC 2000 provides information about the two templates that provide the closest matches to an input utterance. For each of these templates, the matching algorithm yields a single figure that represents a 'goodness-of-fit' between the input and the stored template. In most application software, the decision rule applied to this information involves simply accepting the template with the lower score. Less commonly, the difference between first and second choice scores is used as a basis for assessing confidence in the identification. If this difference falls below a threshold value, the recognizer output might be suppressed, or the user asked for confirmation.

These decision rules have the advantage of simplicity, but are unlikely to be optimal in terms of performance. The immediate output from the pattern-matching algorithm is a goodness of fit score; a unidimensional metric. Because of variability in the user's speech patterns, and in ambient noise levels, this score varies from one occasion to the next. For each vocabulary item, the score values obtained in a series of correct identifications will follow some kind of distribution.

A second characteristic of recognizer function was discussed earlier; when an input is misrecognized (on the basis of a conventional decision rule), the template that gives the second-closest match is usually the correct one. Often, the difference between first- and second-choice scores is rather small; when misrecognitions occur, the decision is often a marginal one.

The template-matching process implemented in recognizers such as the VPC 2000 is thus a paradigm case of decision-making under uncertainty. The first- and second-choice scores obtained in response to a particular input can be thought of as samples from two underlying distributions. By monitoring score values during a period of extended operation, we can build up a picture of these distributions in order to aid the decision-making process in marginal cases.

One important attribute of these score distributions is the central tendency. The vagaries of template elicitation are such that some templates are inevitably 'better' than others; in the sense that, if one were to look at the distributions of scores for the task vocabulary over a period of time, then some templates would tend to produce lower scores than others. Properties such as word length and spectral characteristics may interact with the template matching algorithm, further adding to the variability of score distributions.

It follows that a situation might arise such as that shown in Fig. 13.1. Here the distributions represent the scores obtained in a series of trials when the words A and B are spoken and matched to their respective templates (i.e. these are the scores for correct matches). Suppose we now have an unidentified utterance, that produces the scores SC_a and SC_b when matched to these two templates. Since $SC_a < SC_b$, application of the usual decision rule would lead to the acceptance of word A as the first choice response. However, from the distribution information, it is evident that on the great majority of occasions, when word A is spoken, the score obtained is better than SC_a. On the other hand, a score of SC_b represents a very good score for template B. Given this pair of scores, there is a case for regarding word B as a more plausible response.

An alternative decision rule

Preliminary findings from our experimental data entry task suggest that monitoring of score distributions may be a useful exercise. Suppose that for a given input, the recognizer produces template matches for T_a and T_b, with scores SC_a and SC_b. If we have a history of scores for correct recognitions of these items, we can then express SC_a and SC_b in percentile terms. This value can be regarded as a likelihood

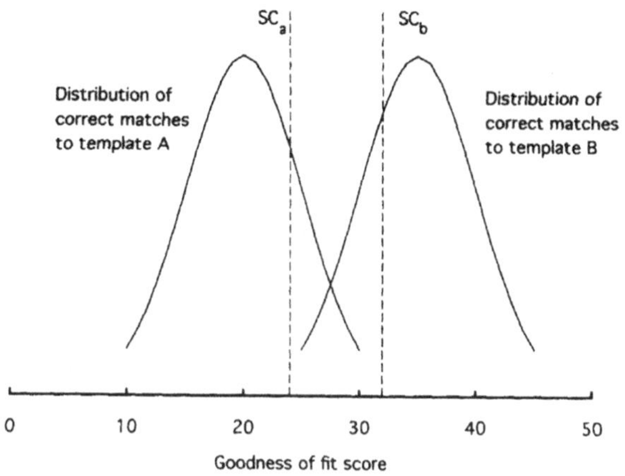

Figure 13.1 Hypothetical distributions of scores for correct recognitions based on two templates, A and B.

estimate. In other words, if there were only 25 per cent of successful matches to T_a with matching scores as poor as or worse than SC_a, the percentile score in this case would be 25.

We can now formulate two alternative decision rules:

- accept the match which is closest in absolute terms;
- accept the match which is more 'typical' – i.e. the one with the higher percentile score.

Preliminary findings from our experimental data entry task suggest that this approach may be a fruitful one. Data from an experimental run were re-analysed, using a modified decision rule based on the principles outlined above. After initial template training, the 'first past the post' rule was applied in all cases. Score data were logged, however, to build up a picture of score distributions as the task progressed. For each input, both decision rules were applied to the score data for first- and second-closest matches. In the event of disagreement, the rule selected was the one that had most often been successful over the set of preceding trials.

The effects of this decision-making strategy are shown in Table 13.2. These data were obtained from a total of 16 subjects, who completed a 30-minute session of the data entry task, beginning immediately after initial template training. During the session, subjects entered around 750 digits, so that by the end of the session, score data were available for about 75 successful recognitions of each vocabulary

Table 13.2 The effects of adaptive decision making on % recognition errors

Condition	Period of task 1	2	3	4
No addaptive control ('first past the post')	3.12	3.72	4.98	4.83
With adaptive control	3.18	3.18	4.15	3.73

item. For the purpose of analysis, the session has been divided into four epochs. The graph shows the percentage of recognition errors as a function of time on task, for the normal 'first past the post' and also for the 'adaptive' decision rule. Both show an increase in error rates with increasing time on task; a pattern which is generally observed (Frankish *et al.*, 1992). However, there are also indications that the use of statistical information about score distributions can enhance recognition accuracy. These data were analysed using a two factor ANOVA, which revealed significant main effects of decision rule and time on task ($F(1,15)=16.61$, $p<.01$; $F(3,45)=4.19$, $p<.05$). There was also a significant interaction between these two factors ($F(3,45)=6.24$, $p<.01$).

The nature of the interaction is evident from Table 13.2; as more information is accumulated about recognition outcomes, the adaptive decision rule shows a progressively larger advantage. In the final task period, the mean error rates for the normal and adaptive systems are 4.83 per cent and 3.73 per cent, respectively. This means that by this stage almost a quarter of recognition errors have been eliminated, largely counteracting the decline in accuracy with increasing time on task.

Some possible limitations; designing for human and system errors

Use of the procedures discussed here is predicated on the assumption that the speech recognition system receives accurate feedback from the user. This will not always be the case, and some consideration must be given to the consequences of actions that are contingent on erroneous feedback. This is particularly true of adaptive systems, where the consequences of such an event may extend beyond the boundaries of the current transaction.

The two types of error that are most likely to be encountered are:

- human error – either failure to monitor feedback correctly, or procedural errors;
- system errors – specifically, cases where a correct recognition is followed by a misrecognition that causes the system to enter an error-correction routine.

The sources of these two types of error are discussed in greater detail by Frankish and Noyes (1990). They found that the commonest of these events were failures to monitor feedback, and that this is more likely to occur when visual rather than spoken feedback is used. In the experimental task described here, subjects failed to correct 17 per cent of recognition errors when feedback was visual; less than 1 per cent when spoken feedback was used.

Whatever their cause, the consequence of any human and system errors will either be acceptance of an incorrect recognition, or rejection of a correct one. How would these events affect the operation of the kinds of system discussed here?

Acceptance of an incorrect entry has no implications for the operation of error-correction dialogues, since these are not brought into play. In an adaptive system, these events will cause erroneous data to be added to the database that underpins the adaptive process. Although this is undesirable, it should not pose too much of a problem in practice. Feedback monitoring failures are most likely to occur when recognition accuracy is high. Even allowing for failures of feedback monitoring, the recognition database built up in the experimental task described here should contain fewer than 1 per cent errors.

Rejection of a correct entry is more problematic. This kind of event can occur as a result of human error, or when the recognition system incorrectly interprets an input as a command phrase such as *try again*, which is used to initiate error-correction. The suggestion was made earlier that rejection of a recognition attempt might cause the corresponding template to be disabled during the subsequent error-correction procedure. This would make it impossible to correct the error by simple re-entry. Because this potential difficulty exists, error-correction dialogues should be designed so that recovery is possible under these circumstances.

In summary, the evidence so far available suggests that although feedback obtained by the system from the user is not entirely error-free, the inaccuracies that do arise are unlikely to be critical.

Conclusions

In monitoring and correcting the system's recognition attempts, the user supplies information that can be of considerable utility. This information may be specific, in the form of a record of attempts to identify the current speech input. It may also be statistical, detailing the characteristics and regularities of recognition performance for each vocabulary item over an extended period of use.

None of the procedures discussed here is particularly profound; more effective strategies of both types can doubtless be devised. Their function here is to illustrate the general design rule that feedback options should be considered in relation to their possible usefulness as a source of information to the system, as well as to the user. These examples can be regarded as specific instances of a general principle; namely, that a well-designed system will exploit every available source of information in the recognition process. The current generation of speech recognizers are highly efficient signal-processing and pattern-matching devices. With a co-operative user, their performance must approach that of humans, in terms of context-free recognition. Human recognition of isolated words taken from connected speech is extremely poor; accuracy is typically around 50 per cent (Pollack and Pickett, 1964). Humans are, of course, superbly efficient at processing contextual information, and use many kinds of linguistic and situational knowledge in speech decoding. This is in sharp contrast with many (but not all) voice input systems that have highly efficient signal processing resources, but a very poorly developed knowledge-based contribution to the recognition process. This kind of contribution does not require fully developed AI software support, although this may be a long-term goal. The message here is that quite rudimentary software refinements can exploit feedback information to achieve worthwhile improvements in recognition accuracy.

References

Damper, R. I. and McDonald, S. L., 1984, Template adaptation in speech recognition, *Proceedings of the Institute of Acoustics*, **6,** 293–300.

Frankish, C., 1987, Voice input applications and human factors criteria, *Proceedings of International Speech Tech '87*, NY: Media Dimensions, 133–6.

Frankish, C., Jones, D. and Hapeshi, K., 1992, Decline of speech recognizer performance with time: Fatigue or voice drift?, *International Journal of Man–Machine Studies*, **36,** 797–816.

Frankish, C. and Noyes, J., 1990, Sources of human error in data entry tasks using speech input, *Human Factors*, **32,** 697–716.

Graffunder, K., Phillips, C., North, R. and Harman, C., 1985, Dynamic retraining of airborne voice recognizers, *Honeywell Systems and Research Centre, Report No. 85SRC31*.

Green, T. R. G., Payne, S. J., Morrison, D. L. and Shaw, A. C., 1983, Friendly interfacing to simple speech recognizers, *Behavior Information and Technology*, **2,** 23–38.

McInnes, F. R., Jack, M. A. and Laver, J., 1987, Experiments with template adaptation in an isolated word recognition system, in J. Laver and M. A. Jack (eds), *European Conference on Speech Technology*, **2,** 484–7.

Martin, T. B. and Welch, J. R., 1980, Practical speech recognizers and some performance effectiveness parameters, in W. A. Lea (Ed.), *Trends in Speech Recognition* 24–38, Englewood Cliffs, NJ: Prentice-Hall.

Noyes, J. M. and Frankish, C. R., 1987, Voice recognition: Where are the end-users? in J. Laver and M. A. Jacks (Eds), *Proceedings of the European Conference on Speech Technology*, vol. 2, 349–52, Edinburgh: CEP Consultants.

Nusbaum, H. C. and Pisoni, D. B., 1986, Human factors issues for the next generation of speech recognition systems, *Proceedings of Speech Tech '86, Voice I/O Applications Show and Conference, New York*, 140–4.

Pollack, I. and Pickett, J. M., 1964, Intelligibility of excerpts from fluent speech: auditory vs. structural content, *Journal of Verbal Learning and Verbal behavior*, **3,** 79–84.

Schurick, J. M., Williges, B. H. and Maynard, J. F., 1985, User feedback requirements with automatic speech recognition, *Ergonomics*, **28,** 1543–55.

Simpson, C. A., McCauley, M. E., Roland, E. F., Ruth, J. C. and Williges, B. H., 1985, System design for speech recognition and generation, *Human Factors*, **27,** 115–41.

Talbot, M., 1987, Adapting to the speaker in automatic speech recognition, *International Journal of Man–Machine Studies*, **27,** 449–57.

14

Comparing error correction strategies in speech recognition systems

W. A. Ainsworth and S. R. Pratt

Abstract

In a noisy environment speech recognizers make mistakes. In order that these errors can be detected the system can synthesize the word recognized and the user can respond by saying 'correction' when the word was not recognized correctly. The mistake can then be corrected.

Two error correcting strategies have been investigted. In one, repetition with elimination, when a mistake has been detected the system eliminates its last response from the active vocabulary and then the user repeats the word that has been misrecognized. In the other, elimination without repetition, the system suggests the next most likely word based on the output of its pattern matching algorithm. It was found that the former strategy, with the user repeating the word, required less trials.

Introduction

One of the reasons why speech recognizers are not often used as a way of entering data into computers is that they make mistakes. They make mistakes because of the inherent variation in speech (a word is never pronounced twice in exactly the same way even by the same person) and because of interference from noise which distorts the signal. Although current speech recognizers based on hidden Markov models attempt to capture the inherent variations and noise compensation algorithms can be employed to reduce the effects of noise, errors still occur.

In order to produce a usable system it is necessary to provide some means of detecting when an error occurs and then correcting it. Errors can be detected by the user if feedback is provided. This may be auditory or visual: the word recognized may be spoken or it may appear on a screen The user can then respond by saying 'no' or 'correction' when it is wrong. Visual feedback is fast, but in many situations where speech input is appropriate, such as when the hands and eyes of the user are occupied with other tasks, auditory output is required.

One situation where speech input is appropriate is dialling a carphone whilst driving. The hands are occupied with controlling the vehicle and the eyes are busy watching the road, but the voice of the driver can be used to speak the number or name of the person he wishes to communicate with. In this situation auditory feedback should be provided so that the driver can check that no recognition errors have been made.

Some experiments have been performed to evaluate various strategies which might be used to correct recognition errors in this situation.

Strategies

The simplest strategy is to have the user repeat a word when it is misrecognized. However, with this strategy there is no guarantee the recognizer will not make the same mistake again. A better strategy is to eliminate the mistaken response word from the active vocabulary before the user tries again. This way, with a finite vocabulary, the correct response will eventually occur. Normally it might be expected that confusable words will quickly be eliminated and the correct response will occur after one or two iterations. This will be called the 'repetition with elimination' strategy.

Most speech recognizers work by computing the difference between the sequence of input feature vectors and a set of stored feature patterns or, in the case of HMM recognizers, by computing the probability that the input feature vectors were generated by each of the models. These distances or probabilities can be used to rank the words of the vocabulary in the order of the likelihood that they formed the spoken input. Another error correction strategy is to use this information by getting the system to suggest the next most likely word when an error is detected. As the user does not have to repeat the word this will be called the 'elimination without repetition' strategy.

Experiments

In order to compare the two strategies some experiments were carried out to simulate 'voice dialling' of telephone numbers in a car. A speech input/output system consisting of a speech recognition and synthesis board (Loughborough Sound Images, 1988) installed in a PC-AT was used. The recognizer consisted of a 16-channel digital filter bank and a dynamic time warping (DTW) algorighm (Sakoe and Chiba, 1978). The synthesizer was a digital implementation of the synthesis-by-rule system of Holmes *et al.* (1964) and a digital formant synthesizer (Quarmby and Holmes, 1984).

The experiments took place in a laboratory with car noises played in the background. Two conditions were employed for training the recognizer: silence and in the presence of car engine idling noise. Testing took place in the presence of noise recorded in a car driven along main roads at various speeds. The average sound level was set to approximate that in the car during the recording session.

The object of the experiment was to 'dial' four 14-digit numbers by voice, correcting recognition errors by the repetition with elimination strategy. The output of the recognizer (the scores corresponding to the total cumulative distances between the test utterance and each of the reference utterances) was recorded after each trial so that the effect of employing the elimination without repetition strategy could also be evaluated.

The recognizer was used in isolated word, speaker dependent mode. Each subject trained the recognition system by speaking words when prompted. There were two training conditions: silence and engine idling. The vocabulary consisted of the digits (*oh*, *one*, ... *nine*) *hundred*, *thousand*, *double*, *treble*, *correction*, *yes* and *no*.

After training each subject was asked to read four 14-digit numbers into a microphone. After each utterance, the synthesizer repeated the word that had been recognized. If the response was correct the subject spoke the next word on the list, but if the recognizer made a mistake the subject said *correction* and repeated the last word. The synthesizer then asked *Was it* ...? to which the user replied *yes* or *no* as appropriate.

Eight volunteers (six male and two female) took part in the experiment.

Results

The average recognition scores, with no attempt at correction, for each of the subjects for each of the conditions is shown in Table 14.1. It can be seen that, as expected, recognition scores were higher with training in idling noise (85.3%) than with training in silence (80.4%). Chi-squared tests showed that this difference was significant at the 0.1 per cent level.

From the point of view of the user of a speech recognition system, the number of corrections required to attain error-free recognition is perhaps more important than the recognition score as this reflects the total time required to dial a number. This may be expressed as the percentage of extra utterances required. This is shown for each subject in Table 14.2. It can be seen that training in silence required 46.5 per cent more utterances, but only 28.5 per cent more with training in idling noise. Chi-squared tests showed that this difference was significant at the 0.1 per cent level.

When the results were analysed with the elimination without repetition strategy simulated, a similar pattern emerged as also shown in Table 14.2. With training

Table 14.1 Recognition scores (%) for each training condition

Subject	Silence	Idling noise
A	91.8	91.8
B	87.1	85.7
C	88.9	95.0
D	90.0	88.6
E	86.1	86.1
F	86.8	86.8
G	63.6	74.3
H	48.8	73.9
Mean	80.4	85.3

Table 14.2 Percentage of corrections required for each subject

Strategy	Repetition with elimination	Repetition with elimination	Elimination without repetition	Elimination without repetition
Subject	Silence	Idling noise	Silence	Idling noise
A	12.5	10.4	23.6	16.1
B	25.0	30.1	26.1	22.5
C	20.0	6.1	20.7	10.7
D	17.5	20.7	14.3	31.1
E	34.6	30.0	32.5	30.4
F	19.3	19.3	32.5	36.1
G	103.6	51.4	110.4	50.4
H	139.3	60.4	193.6	81.1
Mean	46.5	28.5	56.7	34.8

in silence 56.7 per cent more utterances were required whereas after training in idling noise 34.8 per cent more utterances were needed. Chi-squared tests showed that this difference was significant at the 0.1 per cent level.

Comparing the percentage of corrections required by the repetition with elimination strategy with the elimination without repetition strategy (Table 14.2) shows the former strategy to be superior. Chi-squared tests showed that the difference between strategies was significant at the 0.1 per cent level for both training conditions.

Discussion

It has been found that recognition errors are corrected more readily if the user repeats the word that was misrecognized rather than allowing the system to guess on the basis of pattern matching differences. There are two possible explanations of this. An error may be caused by abnormal pronunciation, but when the user repeats the word he is likely to be more careful. Alternatively an error may be caused by the speech signal being masked by noise. Repeating the word gives the possibility of the word being spoken during a period when the background is less noisy, leading to correct recognition.

Despite the wide variation in the recognition scores of the subjects the results indicate that it is better to train a speech recognizer in engine idling noise than in silence. This is in accord with the findings of Kersten (1982) who found that it was better to train a recognizer in the environment in which it is to be used than in silence. In the case of a car environment it is impractical to train a speech recognizer whilst driving, but it is preferable to train it with the car stationary and the engine idling rather than with the engine turned off.

Conclusions

It has been found that fewer corrections are required with a repetition with elimination strategy than with an elimination without repetition strategy.

Acknowledgements

We are indebted to the staff and students of Keele University who acted as subjects in the experiments. The work was supported by EC ESPRIT contract 2101 'Adverse-environment Recognition of Speech'.

References

Holmes, J. N., Mattingly, I. G. and Shearme, J. N., 1964, Speech synthesis by rule, *Language and Speech*, **7**, 127–43.

Kersten, Z. A., 1982, An evaluation of automatic speech recognition under three ambient noise conditions, in Pallett, D. (ed.), *Proceedings of the Workshop on Standardization of Speech I/O Technology*, National Bureau of Standards: Gaithersburg, MD.

Loughborough Sound Images, 1988, uPD7763/4 PC Card for Speech Recognition and Synthesis, Issue 3.

Quarmby, D. J. and Holmes, J. N., 1984, Implementation of a parallel-format speech synthesizer using a single-chip programmable signal processor, *IEE Proceedings*, **131**, (F 6), 563–9.

Sakoe, H. and Chiba, S., 1978, Dynamic programming algorithms optimization for spoken word recognition, *IEEE Trans., ASSP-26*, 43–9.

15

Data – entry by voice: facilitating correction of misrecognitions

A. C. Murray, C. R. Frankish and D. M. Jones

Abstract

Despite technological advances automatic speech recognition (ASR) is still prone to recognition errors. Consecutive misrecognitions during a single attempt to enter a data-item by voice are clearly undesirable. This experiment evaluated three different methods of increasing the likelihood of success when re-entering an item immediately after a misrecognition. The three methods, which constituted the three experimental conditions, were: template removal with immediate re-instatement (IR), serial removal of templates (SR) and *n*th choice (NC). Subjects entered data items by voice under each of the three conditions; half received both visual and verbal feedback during the task, the remainder visual feedback only. For the highly confusable vocabulary used, the procedure NC proved most efficient in terms of the number of successful corrections effected and the speed with which they were completed.

Introduction and background

ASR devices capable of some degree of speaker-independent connected-word recognition are already becoming generally available, but the ASR technology currently available at the more inexpensive end of the market still uses template matching of isolated words (or very limited connected speech) and is usually speaker dependent. Despite the limitations, there are viable commercial applications for this technology, one of which is in the area of data-entry tasks.

ASR devices are becoming more robust against recognition errors and various means can be employed within the particular application to minimize their occurrence, but misrecognitions cannot as yet be eliminated especially if the task vocabulary includes easily confused words. When a misrecognition does occur, attempts to re-enter the correct item may again meet with the same misrecogntion. Clearly such consecutive misrecognitions of a single input are to be avoided: they disrupt dialogue flow, cause frustration and may result in a loss of confidence in the system. Care must be taken in designing the structure of the human–computer dialogue (Simpson *et al.*, 1985, Jones *et al.*, 1989, Molich and Nielsen, 1990): it should help the user detect misrecognitions, and provide consistent, minimally disruptive methods of correcting errors. Error-trapping mechanisms may be implemented at different levels of dialogue processing. This chapter discussed mechanisms appropriate for implementation at a very localized level: they are intended to help resolve misrecognitions still occurring amongst template sets which have already been constrained by higher level error-prevention procedures.

Previous work on this level of ASR interfaces has tended to focus on aspects of the interaction itself (e.g. human factors aspects of dialogue design (Gaines and Shaw, 1984), the timing of feedback (Schurick *et al.*, 1985), error-correction dialogues (Frankish, 1989)) rather than on system-based methods of reducing the likelihood of recurrent recognition errors. The aim of this study was to implement three such methods and compare their efficiency in terms of success rate, speed and ease of use.

Two methods (SR and IR) involved template removal: when a recognition was rejected by the user as being incorrect, the template for the recognized word was removed from the current set. Thus when the user repeated the input, the probability of a correct recognition was increased, not only because the template set had been reduced in size but also because the removed template was likely to have been the one most easily confused with the word actually spoken. The difference between these two methods was that in IR the deleted template was restored immediately after the next recognition was effected, whereas in SR the rejected templates were deleted serially until a recognition was effected, the set size decreasing prior to each re-entry.

The third method, NC, did not remove templates, but instead offered the 'next best match' without the user having to repeat the input. This method exploited the fact that when an isolated word ASR device receives input, it is matched to each of the live templates and a 'goodness-of-fit' score is calculated for each of these matches. The result is that for every input the machine produces a hierarchy of possible recognitions ranked according to the closeness with which each matched the input. Thus, when the user rejects a recognition, the system can offer the next choice in the hierarchy without the user having to repeat the input.

Due to the localized level at which these re-entry methods operated, the primary function of the experimental task was simply to elicit instances of re-entries. For this reason a highly confusable vocabulary was chosen to ensure that the re-entry methods under test would be called on to operate frequently.

Method

The subjects were twenty-four students aged between 18 and 25, all native English speakers. Each completed a block of twenty trials in each of the three experimental conditions described below, balanced for order. Each subject was tested individually, the procedure (including practice time) taking approximately one hour to complete.

On each trial, subjects saw displayed on a computer screen two grids of six squares, one in the top half of the screen, and the second exactly below it in the lower half. At the outset of a trial each square in the upper grid contained a letter (in red) drawn from a sub-group of the E-set (namely B, C, D, V, T and P). The squares in the lower grid were blank. The subjects' task was to complete the lower grid as a duplicate of the upper one, using voice input only. The letter recognized at each input was displayed in white at the appropriate position on the lower grid, and could be deleted if incorrect. Once confirmed as correct (by whichever means was applicable to the particular experimental condition), the colour changed to red. When the lower grid was completed, the subject pressed a key to trigger display of the next trial.

The three main experimental conditions were created by manipulating the underlying action taken by the system when the user deleted (i.e. rejected) a recognition. The experimental task in each condition was identical – the difference lay entirely in the re-entry method used, and the dialogue changes which were attendant upon its implementation. The purpose of all three of these methods was to avoid the occurrence of consecutive misrecognitions of the same input: they sought to increase the likelihood of an item being matched to the correct template when it was re-entered after a misrecognition. The three methods used were as follows.

1. **Template removal with immediate reinstatement (IR)** A misrecognition was deleted by the command *No*. On receipt of this command, the deleted item's template was dropped from the 'live' set for the following recognition, but restored immediately after that recognition had taken place. This method depended for its success on the main patterns of confusion occurring between pairs of vocabulary items, assuming that after one member of the confusable pair was eliminated the repeated input would be matched with the correct template. Since removed templates were restored immediately, there was no need for a confirmation command: input of two consecutive letter names (that is, not separated by a deletion command) served to signal to the machine that the user had moved on to the next letter in the sequence.
2. **Serial removal of templates (SR)** The user deleted a misrecognition with the command *No*. On receipt of this command, the template for the deleted item was dropped from the 'live' set as for IR, but was not restored until a confirmation command was given. Thus if there were consecutive misrecognitions of the same input, the size of the live template set diminished with each rejection. A confirmation command was necessary as an instruction to the machine to re-instate the full template set: the command word used was *OK*.
3. **Nth choice (NC)** The user deleted a misrecognition using the command *No*. On receipt of this command, the system immediately offered the second-best match as a substitute, which the subject could then accept or reject. Acceptance was by default – that is the system took entry of a new letter-name input as a signal that the previous offer was correct – although following pilot studies it was modified to accept (though not require) the explicit confirmation *Yes*. (It should also be noted that this dialogue only partially reflected a true '*n*th choice' procedure sinc the ASR device used only gave access to the first and second best matches).

Although it was desirable that there be a high number of first-entry recognition errors, it was equally important to prevent subjects' confidence in the recognizer deteriorating to a point at which their performance on the task would be adversely affected. For this reason a 'ceiling' procedure was implemented which ensured that the third attempt to re-enter an item always met with a corrected response, whether or not it had in fact been correctly recognized.

In addition to the above three conditions, subjects were randomly assigned to one of two groups A and B. Both groups received full visual feedback during the task – that is to say that the recognition of a letter-name was signalled visually by the display of the letter on the lower grid, and on receipt of a deletion command the most recently recognized letter was removed from the display. However Group A received additional auditory feedback in the form of pre-recorded natural voice messages, which echoed the name of the letter recognized, confirmed deletion, and prompted the user when appropriate to *try again*. In the NC condition, the voice recording accompanying the offer of a next-best match carried a rising (questioning) intonation. Group B did receive some auditory signals: in both groups the system indicated its readiness for voice input with a beep, and the *finished* command (which completed a trial) was double-checked via a verbal message.

Results and Discussion

Details of each transaction between the subject and the system (template matches, distance scores, system reponses and timings) were recorded automatically during the experiment. The spoken input and any departures from the prescribed dialogue were transcribed and checked against audiotapes recorded during the sessions.

Measures of error correction success

Re-entry success rate was measured for the two groups combined. The number of misrecognitions was counted, and the number of re-entry attempts made for each of these misrecognitions was noted. Figure 15.1 shows for each condition the mean percentage of errors corrected at the first and second attempts combined, and Fig. 15.2 shows how these corrections were distributed between the first and second attempts.

A one-way related ANOVA showed the differences in error-correction rates seen in Fig. 15.1 to be significant ($F(2,46)=18.47$, $p<0.0001$). *Post-hoc* t-tests showed each of the three conditions differed from one another significantly (between IR and SR, $t(23)=2.58$, $p<0.01$; between SR and NC, $t(23)=3.94$, $p<0.005$). From Fig. 15.2, it can be seen that at the first attempt at re-entry of a misrecognized item, IR and SR performed equally well: this was to be expected since at the first attempt at correction, both these methods had dropped one template from the live set, and it was only at the second attempt that the two methods diverged. In NC, significantly fewer re-entries were successful at this first attempt (related ANOVA:

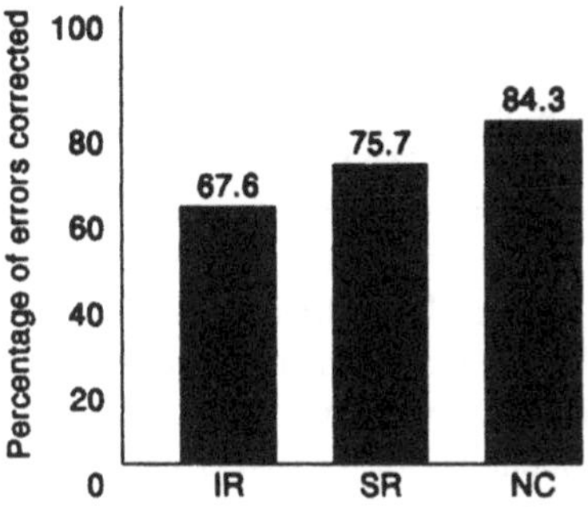

Figure 15.1 % Errors corrected: Total for 1st & 2nd attempts

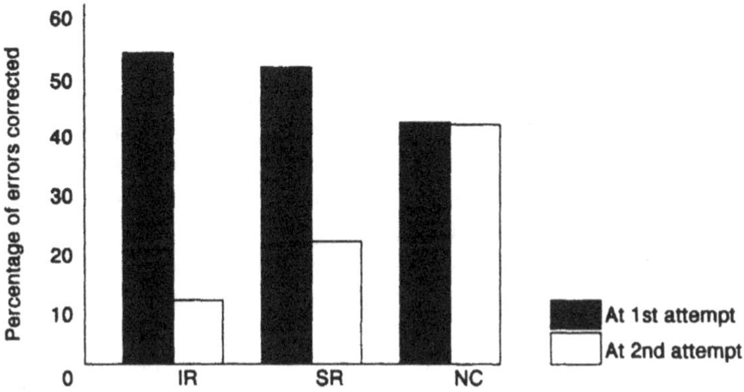

Figure 15.2 % Recognition errors corrected at 1st & 2nd attempts

$F(2,46)=7.2, p<0.002$). It seems that, at least with the particular vocabulary used, the second-best template match (as offered by NC) was a less reliable reflection of the input than a recognition effected by its repetition (as in IR and SR). At the second attempt at re-entry of a misrecognized item, IR performed least well, SR (which at this stage had two templates deleted from the live set) performing better and NC being the most effective. The percentage unaccounted for in the figures represents those entries which remained uncorrected at the first and second attempts, and thus reached the 'third-attempt ceiling' described above (32.4 per cent in IR, 24.3 per cent in SR and 15.7 per cent in NC).

It should of course be borne in mind that these results reflect performance of the re-entry methods with regard to the particular, highly confusable vocabulary selected here. Where the vocabulary is less thoroughly confusable so that misrecognitions are more likely to occur between particular pairs of words rather than between whole groups of words, Fig. 15.2 indicates that IR or SR will be the more effective methods. This illustrates the importance of investigating the confusion patterns inherent in the vocabulary and commands selected for an application before implementing a particular re-entry method.

Time measures and the efffect of auditory feedback

The time taken to complete an error-free entry under each of the conditions was recorded, together with the time taken to effect a correction successfully at the first attempt. The difference between these measures yielded a time-overhead for effecting error corrections in each of the conditions. The results conformed with expectations generated by the differences in the dialogue structures.

All entries in Group A took longer than in Group B, because of the extra time taken up by the spoken feedback given to Group A. Apart from this overall difference, the pattern of results was identical for the two groups. The mean times (in seconds) to complete one error-free input are shown in Table 15.1. Within each group, there was no difference between the times required in IR and NC to effect a single entry, but in SR this took significantly longer due to the need to enter the confirmation command 'OK' which was not necessary in the other conditions ($F(2,22) = 196.94$ (Group A) and 244.27 (Group B), both $p < 0.0001$).

Table 15.2 shows the mean time-overheads (in seconds) to effect an error correction in each of the conditions for the two groups. Error correction in IR and SR carried an equal time-overhead because each required a deletion command and re-entry of the item, but the time required in NC is significantly shorter because the input did not have to be repeated in order to effect the correcion ($F(2,22) = 57.76$ (Group A) and 74.11 (Group B), both $p < 0.0001$).

The absolute time values are of course directly related to the particular vocabulary and dialogue selected. These results demonstrate that the relationship between data-entry dialogue structure and the times taken to complete entries and corrections is quite straightforward, so that it is possible to make accurate predictions about the time course of sections of the interaction based solely on information about its structure and single-word input-times.

However in order to evaluate the overall time efficiency of alternative potential data-entry dialogues, it is also necessary to know how variations in their structure are likely to affect the ease with which people are able to use them. The inclusion of auditory feedback is one way in which the dialogue structure might be varied, and the intention here was to investigate whether or not subjects were assisted by its presence and whether any advantages found would be sufficient to offset the

Table 15.1 Mean time (secs) for completion of single error-free entry

	IR	SR	NC
Group A	1.66	2.84	1.71
Group B	1.45	2.52	1.47

Table 15.2 Mean time-overhead (secs) for effecting a successful correction

	IR	SR	NC
Group A	3.27	3.14	2.18
Group B	2.60	2.53	1.50

considerable time overhead involved. Subjects' dialogue errors were classified in an attempt to make comparisons but unfortunately the numbers of errors were isufficient to allow any clear conclusions to be drawn. However, it is worth noting that where they did occur, most of these errors were associated with a particular dialogue (IR, SR or NC) rather than with a particular feedback mode (A or B). There was one exception to this: errors of omission occurred in Group B but were rare in Group A. A particular example of omission in Group B subjects was the failure to enter the deletion/rejection command *No* in the NC dialogue: subjects just repeated the correct input without rejecting it first. It may be that the provision of spoken feedback to Group A made the rhythmicity of turn-taking more obvious than it appeared when the machine was silent (as it was for Group B).

Conclusions

The experiment described used a simple voice data-entry task to investigate three different methods of increasing the likelihood of a correct recognition on re-entry of a misrecognized word.

For a small highly confusable vocabulary such as the one used here, an '*n* choice' method was the most successful in terms of the time taken to correct a misrecognized item, and the number of successful re-entries effected. An '*n*th choice' procedure is also simple to implement as it does not involve template removal, does not require a confirmation command, and the user does not have to repeat the input (a factor which may save voice fatigue over long periods) - the only requirement is for a deletion command. However repetition of a word met with a higher success rate than simply taking the second best match to the initial input. This indicates that where confusion patterns lie between particular pairs of items in a vocabulary (so that re-entry is likely to succeed as soon as the 'wrong' member of the pair has been removed), template removal with immediate re-instatement may provide better re-entry accuracy without incurring a large time overhead.

The provision of verbal feedback messages may help users to make the correct dialogue moves by making the dialogue seem more natural, but carries a high time cost. It may only be desirable to include verbal messages in data-entry tasks where users cannot give full attention to the visual display and need auditory responses to prompt and guide them through the dialogue.

Acknowledgement

This work has been carried out with the support of the DRA.

References

Frankish, C., 1989, conversations with computers: problems of feedback and error correction. *Proceedings of European Conference on Speech Technology*, **2**, 589–92.

Gaines, B. and Shaw, M. G., 1984, *The Art of Computer Conversation*. Prentice Hall.
Jones, D., Hapeshi, K. and Frankish, C., 1989, Design guidelines for speech recognition interfaces, *Applied Ergonomics*, **20**, 47–52.
Molich, R. and Nielsen, J., 1990, Improving a human–computer dialogue, *communications of the ACM*, **33**, 338–48.
Schurick, J., Williges, B. and Maynard, J., 1985, User feedback requirements with automatic speech recognition, *Ergonomics*, **28**, 1543–55.
Simpson, D., McCauley, M., Roland, E., Ruth, J. and Williges, B., 1985, System design for speech recognition and generation, *Human Factors*, **27**, 115–41.

Part IV: Designing and Evaluating Interactive Speech Technology

16

Designing and evaluating interactive speech technology

C. Baber

Introduction

In this final section we consider the issues involved with designing interactive speech technology. Underlying this area of work are questions of how to ensure that the speech system we have designed will work effectively and, of equal importance, actually be used. With reference to this latter point, if potential users of this new technology do not find it attractive, they are unlikely to use it; all the effort spent on designing an interactive speech system is then to no avail.

Chapter 17, from Cowley and Jones, considers the relationship between the quality of synthetic speech and listeners' preferences. It is fairly well established that the quality of synthetic speech has marked effects on the manner in which people will respond to it. Although we all use the term 'quality' to refer to synthetic speech, it has proved remarkably resilient to definition. One approach to developing a useful definition of speech quality is through the statistical practice of factor analysis. Cowley and Jones' study, involving a number of subjects and several speech systems, produced two principle factors: listenability, i.e., subjects' aesthetic response to the voice, and assertiveness/clarity, i.e., the manner in which the voice conveys information. These factors are somewhat different to those usually associated with the use of synthetic speech, such as intelligibility, and the authors provide an interesting discussion of their results.

I noted in the first chapter in this volume that we require practices and methods which will allow us to determine how people will communicate with interactive speech technology. There have been a number of studies employing the Wizard of Oz (WoZ) paradigm, i.e., where an experimenter simulates the performance of either a speech synthesis or recognition device, or, more common, in interactive speech system. This paradigm is comprehensively reviewed by Fraser and Gilbert (1991). In general the structure of the WoZ interaction is controlled by an experimenter, which may lead to some inconsistency in the performance. Further, data is normally collected after the experiment, either from a transcription of the interaction, and from coding of key elements. Both of these aspects can lead to problems, and have been dealt with in Chapters 18 and 19. Tate *et al.* have developed a methodology for the rapid prototyping of interactive speech dialogues, based around a WoZ principle. The dialogue controller software allows real time data collection from interactions. These data can be used in evaluation exercises, and can lead to the redesign of dialogues.

Foster *et al.* discuss a variation on WoZ. Dialogue control is exercised via JIM (Just sIMulation), which creates a path through a dialogue space to provide prompts to the user. The experimenter responds using a keyboard, and JIM converts the keyed response into a spoken message. The details of the interaction are also logged in real time. In addition to collecting objective data concerning time taken, number of errors, etc., both of the papers also make use of subjective data, asking users about their perceptions of the system and whether they thought it could be improved. User centred design has had a lengthy and productive history in ergonomics, and more recently, in the field of human–computer interaction (Norman and Draper, 1984).

Chapter 20, from Hapeshi, presents a set of eight guidelines for the use of speech in multimedia systems. The rapid development of computer technology has made the use of a range of information presentation media possible (Lewis, in Chapter 4, has discussed the potential role of speech in CAL). Speech can provide a means of broadening the bandwidth for information presented to the user; which can enhance the interaction, or simply confuse and annoy the user. Thus, it is important to have some idea of when to use speech. Hapeshi's guidelines are presented in the context of multimedia systems, but ought to be considered by all of us involved in the design of interactive speech systems.

References

Fraser, N. and Gilbert, G. N., 1991, Simulating speech systems, *Computer Speech and Language*, **5**, 81–99.

Norman, D. A. and Draper, S. W., 1984, *User Centred Design* [Hillsdale, NJ: LEA].

17

Assessing the quality of synthetic speech

C. K. Cowley and D. M. Jones

Abstract

In human–computer interaction, users' perceptions of speech style and quality influence their use of and reactions to the whole system. Systematic and anecdotal evidence suggests that even high quality synthetic voices have the potential to become irritating to listeners, regardless of their intelligibility.

In order to determine the relationship between voice qualities and listeners' preferences, a wide range of synthetic voices were rated on thirty perceptual scales by forty subjects. Emphasis was on objective evaluation of voice quality rather voice style or speaker characteristics. Factor analysis of the perceptual ratings recovered two strong perceptual factors termed **Listenability** and **Assertiveness/clarity**. These factors and their relationships to specific voice parameters are discussed in the light of previous research.

Introduction

An abundance of speech output devices has been developed for integration into computer systems. Some of these devices, such as the Digital Equipment Corporation's DECtalk and DECvoice, produce speech which is sufficiently high quality to allow implementation in a variety of commercial, industrial, educational, and military applications.

The effefctive implementation of speech technology is a far more complex concern than a simple one-to-one mapping of manual input and visual displays to I/O alternatives in the auditory domain. Users' perceptions of speech quality or style influence their use of and reactions to the whole system in a number of ways. In some cases, efficiency or intelligibility is less important to users than simple preferences for particular speech styles (Rosson and Mellen, 1985). As well as systematic evidence there are numerous anecdotal examples of pronounced (and usually negative) reactions to machine speech. There are reported incidents in which drivers have disabled dashboard voice-warning systems (Peacock, 1984). Such occurrences suggest that an understanding of the variables contributing to such reactions is critical to the success of voice output systems.

Speech conveys much more than linguistic information. Features such as speaking rate, amplitude, pronunciation, accent and dialect enable listeners to draw conclusions about the characteristics of the speaker and make predictions about race, sex, personality, emotional state and credibility. Equally with synthetic speech systems, personality is ascribed on the basis of speech style. It is not uncommon for listeners to describe unfamiliar synthetic voices in negative terms such as 'evil' or 'sinister' (Michaelis and Wiggins, 1982). Synthetic speech can be modified to sound more natural by adding or improving rule-generated stress, but this decreases intelligibility (McPeters and Tharp, 1984). There appears to be a negative correlation between intelligibility of synthesized speech and its aesthetically pleasing characteristics.

Factor analysis and scaling methodologies have been employed in a number of studies which have attempted to discover how listeners' perceptions are influenced by underlying characteristics of speech quality and style. In many cases, researchers have adopted perceptual scales derived from personality research to examine users' assessments of human voices. In 1975, Giles and Powesland used a set of personality ratings in a study of speech style and social evaluation (i.e. hard working/lazy, honest/dishonest, generous/ungenerous, wealthy/poor). Although they found that two factors, 'agreeableness' and 'assertiveness', appeared to be related to judged appropriateness of voices for tasks, it is not clear what perceptual qualities of the speech were responsible for these personality attributions. Despite this, Cox and Cooper (1981) used predictive measures derived from the same set of personality ratings in an attempt to assess the relevance of speech style as an indicator of various speaker attributes. A scaling methodology was used to determine the 'features of the speech that were considered by the subjects important in selecting a preferred speaker'. Unfortunately, the paired comparison methodology used in the study tells us little more than how one voice compares with another on a set of personality scales. This may be useful if a practical decision needs to be made regarding choice amongst a limited voice set for a specific application, but explains little about the specific qualities of the voices that influence listeners' preferences. Both studies tend to overlook the distinction between the personality listeners ascribe to a speaker and their perceptions of the voice's more objective features.

A more exacting and replicable example of the application of factor analysis in speech perception scaling was conducted by Rosson and Cecela (1986) who made a simplifying distinction between **quality** and voice **style**. Voice quality is conceptualized as its inherent timbre, and voice style as a function of the manner in which words are produced, their intonational and durational characteristics. Whilst both sets of characteristics will influence listeners' reactions to a voice, an attempt to assess quality independently of style is likely to lead to more objective findings. The use of a synthesizer allows such an assessment because computer speech technology has advanced to the point where it is possible to directly and precisely manipulate many of the factors which may contribute to the voice quality perceived (such as fundamental frequency and formant spacing). The quality of voice can be modified dramatically with no efffect on such factors as intonation contours, duration and word stress.

Rosson and Cecela (1986) assessed eight listeners' perceptual and appropriateness evaluations of sixteen variations on one synthetic voice modified across four voice quality parameters, richness, head size, average pitch and smoothness (see procedure for descriptions of these parameters). One simple sentence was used for all presentations. Factor analytic and regression techniques were used to map the relationships between voice quality, perceptual evaluations and appropriateness measures. Varimax orthogonal rotation of the data matrices yielded a two factor solution accounting for 76.6 per cent of the total variance. Factor one was characterized as a 'bigness' or 'fullness' dimension with the strongest contributing scales being: big/small, low/high pitch, rich/thin and heavy/light. Regression analyses using the voice quality parameters as predictors showed a strong contribution from the richness variable to 'fullness' with voices high in richness having high scores on 'fullness'. Head size and pitch were also associated, and their relationship with 'fullness' was especially strong when high head size was associated with a low pitch. Factor two was characterized as 'clarity' with the strongest contributors being: clear/muffled, polished/sloppy and smooth/rough. The smoothness parameter had the largest effect on 'clarity' with voices high in smoothness having high 'clarity' scores. Voices high in both 'fullness' and 'clarity' were judged appropriate for 'information provision' scenarios.

Factor analytic studies are constrained by the stimuli and response measures offered to subjects. Rosson and Cecela report that there may well have been important voice quality manipulations and perceptual variables that they failed to include. The current study seeks to address such limitations by the use of a much wider selection of voices, scales and an increased number of judges so that a much more comprehensive evaluation of the issues of speech perception scaling is achieved.

In expanding the voice set, it seemed appropriate to capitalize on the strong effects reported by Rosson and Cecela by using the same parameters in this study. However, this time the modifications were applied to the full range of voices produced by a DECtalk synthesizer. Also, the stimuli were expanded to form a large set of sentences designed to demonstrate the full range of the synthesizer's capabilities. Furthermore, a larger selection of perceptual scales was used and the validity of the ratings obtained was enhanced by the use of a comparatively much greater number of judges.

Procedure

Voice parameter modification

A DECtalk formant synthesizer was used for the experiment. This device, which converts ASCII code into high-quality synthetic speech, is widely reported to be the most superior synthesizer available in terms of intelligibility (Greene *et al.*, 1986) and naturalness (Nusbaum *et al.*, 1984). The device was controlled by a MicroVax and a standard VT340 keyboard and monitor was used for the presentation of the scales and the collection of data from the subjects.

The DECtalk produces seven default voices. The six most distinct were chosen: standard male/female, deep male/female, older male and child. These were then adapted using four design parameters resulting in a set of twenty-four different and distinctive synthetic voices.

Smoothness is caused by a decrease in voicing energy at higher frequencies and is the opposite of brilliance. Professional singing voices trained to sing above an orchestra are usually high in brilliance. The smoothness parameter was used because we would intuitively expect modification in this dimension to produce changes in voice quality but to have marginal influence on the perceived identity of the speaker. For each voice, smoothness was increased by 50 per cent.

Voice **richness** or forte is similar to brilliance and the opposite of smoothness. It appears to be associated with the appearance of a low amplitude nasal formant. Rich voices carry well and are more intelligible in noise environments. Richness was modified by either + or −50 per cent depending on the default setting. As the criterion for adaptation of the speech was to create a wide range of voices and determine the influence of the parameters in general, rather than in a particular direction, a unidirectional modification across the voices was not essential.

Fundamental frequency indicated in hertz (and perceived as voice pitch) has been suggested for indicating the urgency of a message with relative raises in pitch corresponding with greater perceived urgency (Simpson *et al.*, 1984). **Average pitch** was either increased or decreased by 50 Hz depending on the default pitch. It was raised if the default pitch was less than 50 Hz and lowered if a 50 Hz increase would exceed DECtalk's highest possible setting. **Pitch range** was not modified as this can result in distinct changes in the character or style of the speech.

The **head size** variable is realized acoustically through changes in formant positioning and amplitude. Human head size has a strong influence on a person's normal speaking voice. Humans with larger heads tend to have lower, more resonant voices. Voices with enhanced resonance were created by increasing head size by 15 per cent for the six default voices.

A ratings program was written which randomized the order of presentation of voices scales and sentences and then collected the full 720 ratings from each subject (24 voices rated on 30 scales each). On completion of the task, the program created a data matrix for each subject which was suitable for statistical analysis. Thirty sentences of unemotive semantic content were constructed covering the full range of phonemes provided by DECtalk. Many of the sentences were entered to DECtalk phonetically to remove any pronunciation errors and minimize it's American accent. Thirty bipolar, five-point rating scales were constructed. Forty subjects were used in the experiment. None were familiar with speech synthesis systems.

Results

The data matrices were subjected to factorial analysis. Factors were retained if their eigenvalue was at least one. The correlation cut-off point was set at +/−0.5. Varimax orthogonal rotation yielded a three factor solution accounting for 62.5 per cent of the total variance. As the third factor accounted for only 5.2 per cent of the variance not accounted for by factors one and two, it is not discussed here.

Factor 1: 'Listenability'

This accounted for 44.7 per cent of the variance with the following seven scales making strong contributions (in order of highest to lowest correlation with factor):

Scales		*Correlation*	
Dissatisfied	---	Satisfied	(0.79)
Irritating	---	Not irritating ---	(0.78)
Harsh	---	Gentle	(0.69)
Hostile	---	Friendly	(0.67)
Unpleasant	---	Pleasant	(0.65)
Disturbing	---	Restful	(0.63)
Crude	---	Refined	(0.50)

Regression analysis revealed that the Richness parameter correlated with factor one to 0.57 ($p<0.005$), accounting for 33 per cent of the variance. Smoothness correlated 0.53 ($p<0.005$), accounting for 27 per cent. Together, these parameters accounted for 41 per cent of the variance in factor one.

Factor 2: 'Assertiveness/clarity'

This factor accounted for 12.6 per cent of variance not accounted for by factor 1. The following seven scales correlated with this factor:

Scales		*Correlation*	
Calm	---	Anxious	(0.77)
Relaxed	---	Tense	(0.74)
Authoritarian	---	Meek	(0.69)
Clear	---	Confusing	(0.621)
British	---	Foreign	(0.620)
Composed	---	Confused	(0.55)
Knowledgeable	---	Uneducated	(0.52)

Regression analysis revealed the average pitch parameter correlated with the factor to 0.42 ($p<0.005$), accounting for 18 per cent of the variance. Head size correlated 0.42, accounting for 18 per cent. The analysis showed that these parameters correlated so highly that they cannot really be said to make independent contributions to the variance.

Discussion

When the scales associated with factor one are considered, they can be seen to apply to the aesthetic listenability of the voices. Only one scale seems out of place:

dissatisfied/satisfied. We can only assume that subjects regarded this scale as a measurement of how satisfied they were with each voice rather than a measurement of any levels of satisfaction that might have been expressed in the voices themselves.

The richness and smoothness parameters made independent contributions towards explaining the variance in factor one. The strongest overall contribution came from the richness variable. The characteristics of the 'listenability' factor and associated scales were, however, qualitatively different from Rossen and Cecela's reported 'fullness' factor. Furthermore, head size, which contributed to 'fullness' made no significant contribution to 'listenability'. Differences between the isolated factors and other results may reflect the differences between the scale set, the voice set and the range of stimuli. The voice set used in Rosson and Cecela's study consisted of only one default voice, varied with subtle modifications of the parameters. This may have led to many of the stimuli being perceptually indistinct. The current study used six distinct default voices which were modified fairly dramatically. Consequently, the much wider and more varied selection of voices is likely to have resulted in a more conclusive representation of speech quality perception. Furthermore, the stimulus sentences in the current study included instances of the entire phoneme range of English speech. Ratings obtained from such stimuli would be expected to elicit a greater range of perceptual responses than those obtained by confining the stimulus to a single sentence. Finally, the data obtained from forty subjects is likely to result in more robust conclusions than data obtained from only eight.

Rosson and Cecela consider their study to be only a partial mapping of the relationships among these kinds of variables. We consider the current study to build on the findings of such research and, whilst in no way offering a complete analysis of the issues, to expand and enhance the existing knowledge concerning speech perception.

Moving on to factor two, 'assertiveness/clarity'. Voice pitch has been suggested for indicating the urgency of a message (Simpson *et al.*, 1984). Therefore it is appropriate that scales such as calm/anxious and relaxed/tense correlate with changes in average pitch and head size. This suggests that the factor is associated with the extent to which the voices sounded assured and calm or urgent and panicky.

Conclusions: relevance to voice-output systems in general

Clearly the actual value of the ratings obtained are heavily dependent on the synthesizer under test, although we would predict that these findings would apply to any formant synthesizers of comparable quality. Furthermore, as the DECtalk produces such high quality and natural speech, it is likely that the variables that have been found will have direct relevance not just to synthetic speech, but also to natural speech in general.

Synthetic speech was used because it enables accurate and quantifiable modification of the speech. Exact replication or modification of the experiment is possible where 'natural' (or digitized) speech would be problematic. It seems likely that the results would be similar if an equivalent set of digitized voices could be found. Many studies have highlighted differences between synthesized speech and natural speech but these are mainly concerned with assessing issues of cognitive processing of the speech. It should be noted that the technology of speech synthesis is reaching a point where differences between the quality and naturalness of synthesized speech and that of human speech may become negligible. Hence, it seems likely that the findings would apply to human speech output if the objective features could be analysed and modified with comparative precision.

The study gives us some pointers about how to modify synthesized voices to suit potential users. It should be remembered, however, that it is unlikely that any particular type of synthesized voice will be amenable to speech output in all situations. Rather, the combination of task and user characteristics associated with a particular system will dictate the applicability of speech output. Furthermore, the ratings task is an abstract one and we do not know how closely the results we report here will correspond to an environment where users are interacting with real voice-output systems to achieve routine tasks on a day to day basis. Findings from task-appropriateness ratings may indicate the potential reactions of users but do not preclude studies of users interacting with real speech output applications.

References

Cox, A. C. and Cooper, M. B., 1981, Selecting a voice for a specified task: the example of telephone announcements, *Language and Speech*, **24**(3).

Giles, H. and Powesland, P. F., 1975, *Speech style and social evaluation*. London.

Greene, B. G., Logan, J. S. and Pisoni, D. B., 1986, Perception of synthetic speech produced automatically by rule: intelligibility of eight text-to-speech systems, *Behaviour Research Methods, Instruments and Computers*. **18**, 100–7.

McPeters, D. L. and Tharp, A. L., 1984, The interference of rule-generated stress on computer-synthesized speech, *International Journal of Man–Machine Studies*, **20**, 215–26.

Michaelis, P. R. and Wiggins, R. H., 1982, A human factors engineer's introduction to speech synthesizers, in *Directions in Human–Computer Interaction*, A. Badre and B. Schneiderman (Eds), Norwood NJ: Ablex Publishing.

Nusbaum, H. C., Schwab, E. C. and Pisoni, D. B., 1984, Subjective evaluation of synthetic speech: measuring preference, naturalness, and acceptability, *Research on Speech Perception Progress Rep 10*, Bloomington, IN, Speech Res. Lab., Indiana University.

Peacock, G. E., 1984, Humanizing the man/machine interface, *Speech Technology*, **2**, 106–8.

Rosson, M. B. and Cecela, A. J., 1986, Designing a quality voice: an analysis of listeners' reactions to synthetic voices, *Proceedings of CHI'1986*, 192–7.

Rosson, M. B. and Mellon, N. M., 1985, Behavioral issues in speech-based remote information retrieval, in Leon Lerman (Ed.), *Proceedings of the voice I/O systems applications conference '85*. San Francisco: AVOIS.

Simpson, C. A., Marchionda-Frost, K. and Navarro, T. N., 1984, Comparison of voice types for helicopter voice warning systems, SAE Technical Paper Series 841611, *Proceedings of the Third Aerospace Behavioural Engineering Technical Conference, 1984*, SAE Aerospace Congress and Exposition, Warrendale, PA: SAE.

18

Evaluation and prototyping of dialogues for voice applications

Mike Tate, Rebecca Webster and Richard Weeks

Abstract

This chapter concerns the use of rapid prototyping, and evaluation of dialogues for use with speech-based systems and services at BT.

The chapter discusses the dialogue constructor (dc) used at BT. The prototyper was designed at BT's laboratories for the purposes of rapidly designing dialogues. It describes what the prototyper can do, and how it is used in the design of systems employing speech input and output.

The chapter also discusses the importance of prototyping in giving the design team a true representation of a proposed dialogue as opposed to paper-based dialogue design.

The prototyper is also used in the automated collection of data for analysis of actual user behaviour, including recognizer performance and dialogue effectiveness and efficiency. This recorded data is stored and later accessed for analysis.

In addition, the chapter discusses the need for prototyping in the evaluation of dialogues. It describes how speech-based services are evaluated using the prototyper, and the methods used in the evaluation process. This includes the Speech System Evaluation Methodology (SSEM) used in subjective assessment, and measures used in objective assessment. Subjects' interactions with dialogues on the prototyper are also monitored using audio and visual recordings. This gives a complete picture of user-system interaction.

Evaluation and prototyping

Introduction

Prototyping refers to the simulation of an actual product or service. This simulation incorporates features that are important in the final system.

A prototype enables the designer to demonstrate the behaviour of the final system, and how it will behave. The rationale behind prototyping is the idea that theory or paper based design is not enough to allow a perfect match between the needs of a user and the design of the system. Rather, what is needed is a model that enables an actual interaction to occur between the user and the system, and for this interaction to be observed. This approach may be used in all stages of the development cycle and is critical in the development of systems that meet the accepted standards of usability.

There are four primary advantages to the prototyping approach.

1. It facilitates the rapid design of a system for the purposes of evaluation.
2. It allows for an iterative development cycle, since the prototype system may be continuously evaluated and modified until the required level of performance is attained.
3. Valuable user feedback can be obtained in the initial stages of system design and modifications made accordingly. This ensures that a system meets the customers' requirements.
4. The cost of rapid prototyping is relatively small compared with the cost of launching a product that is not usable and does not meet with the users' approval.

The most effective systems are those that employ an iterative design process and which consult most closely with the user. Thus, rapid prototyping can be the key to the overall success of a product or service, and as such it is a valuable and dynamic design tool.

The most functional dialogues are produced through a process of evaluation which identifies key problem areas and enables them to be improved upon in successive cycles of design. Thus prototyping, which can be used at various stages of the development cycle, is a valuable design and evaluation tool for speech dialogues.

Dialogue prototyping

BT Laboratories at Martlesham Heath, Suffolk, have developed a dialogue constructor (dc) for speech-based services which use either speech recognition or Dual Tone Multi-Frequency (DTMF) input. The dc is a C program that runs under MS-DOS, UNIX or VMS and is driven by a special-purpose, textual dialogue description language.

BT's dc can interface to a number of different I/O devices (see Fig. 18.1). It will accept input via a keyboard, or it can use recognized speech input via microphone or telephone handset, or DTMF input via a telephone keypad. The messages can be output via the screen or via text-to-speech (TTS) synthesis, or as digitally recorded messages held on disc. A front end graphical interface has been developed for ease of use.

The dc is used for the purpose of rapidly implementing and evaluating dialogues for speech-based services. In addition, the dc is used in the automatic collection of data about actual user behaviour, which is analysed for the assessment of dialogue effectiveness and efficiency. The data collection facility also provides information on recognizer performance. Each of these three functions will now be dealt with in more detail.

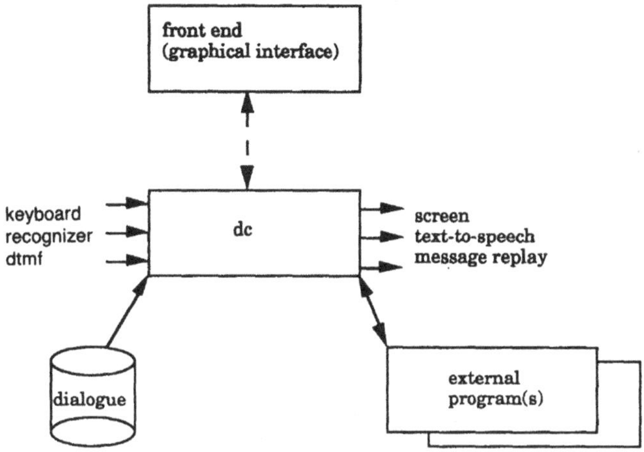

Figure 18.1 Dialogue constructor i/o

Dialogue design

The dc can be used by designers, at various stages of the dialogue design process, to evaluate initial dialogue messages and dialogue flow. This is important since the written word differs from the spoken word, thus a written dialogue may contain inherent problems which only come to light when the dialogue is spoken. This is discussed more fully later on. Thus, rapid prototyping can be invaluable to dialogue designers as it allows them to produce a realistic representation of a dialogue. This simulated dialogue can then be assessed for obvious faults, and the designer does not have to rely solely on a paper-based dialogue design process.

Dialogue evaluation

Rapid prototyping should form an integral part of the evaluation process, as it enables the modification of dialogue messages and structure, and ensures that an acceptable level of performance and satisfaction is attained by users of the system.

Users can interact with a dialogue that is run on the dc, and their performance and reactions towards the dialogue gauged. In addition, several different dialogues may be run in order to determine which one is best suited to the task. The dc is able to simulate different error rates by varying the parameters of a speech recognition model and by varying recognition.

Automated data collection

The dc has the capability to run the dialogue and to log events as well as record the users' utterances. In addition, it records the recognizer's confidence level for the accepted input. This data can then be stored and analysed at a later date. This latter feature is a powerful tool for evaluators, since they would otherwise have to analyse a user-interaction in real time.

The capabilities of the dc, discussed above, result in a rich pool of information with which to analyse a speech-based service.

Evaluation in the design life cycle

Dialogues can be evaluated at different stages of the design cycle; from the very early stages when they are written by the designer, to the point where they are implemented in the final system.

Figure 18.2 shows the dialogue design process. This also shows the stages in the process where evaluation can and should take place, from the initial specification, where the dialogue is compared against design guidelines, to the evaluation of the completed system.

The evaluation of a system after it has been implemented is important. However, at this stage it is often too late to make any major changes to the system as these may be difficult to implement. In addition, changing the dialogue of a completed system may involve suspending a service to customers whilst the changes are made.

Given the above, the evaluation of a speech-based system must take place as far as possible before its final implementation, and this emphasizes the need for prototyping.

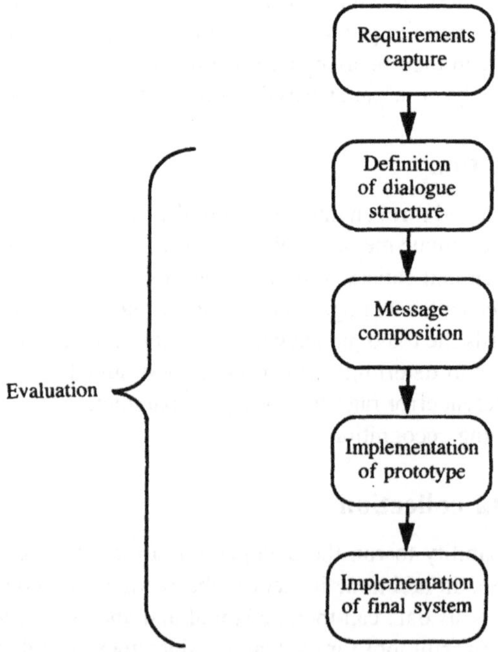

Figure 18.2 The Dialogue Design Cycle.

Evaluation and prototyping

In the construction of dialogues, designers will write the words and phrases of system messages in a dialogue specification. Evaluation takes place at this stage with a paper-based assessment of the system messages. During this stage of evaluation, the comparison of the messages against current dialogue design guidelines, which have been written on the basis of past experience, can be useful (Popay *et al.*, 1991).

The above early evaluation of dialogues is valuable, but it can also be problematic. The use of speech at the interface poses problems in the evaluation of such services on paper in that written dialogues differ from spoken dialogues in three main areas. These differences relate to

- speech as input
- prosody
- accent

Speech as input

Speech information is available to the listener via the auditory modality. It is presented serially. The limitations of short term memory necessarily mean that some of the information in a spoken phrase may be quickly forgotten after it has been presented. In written communication, all the information is available to the reader on the page or on the screen. This allows readers to efficiently scan information, where the stored words on a page act like a form of extended memory for readers to extract relevant information at their leisure. This scanning ability is not available to listeners of the spoken word in the same way. Therefore, the presentation of spoken information must take account of these human constraints.

Prosody

Prosodic information consists of pitch, duration, and intensity. This information is not usually given in written form, but even when it is available in some coded form it is hard to translate in an efficient and effective manner. Yet prosody has an important part to play in human speech communication. The pitch and intensity with which a phrase is spoken can tell us much about the speaker's feelings on what has just been said. In a similar way, speech-based systems will use prosodic information in order to elicit the correct affective responses from users.

The design of different speech-based services, intended for varying sets of users, will employ different prosodic patterns in order to create different images in the mind of the listener. For example, a telephone banking service will want to create an image of efficiency. In contrast, a tourist information service may place the emphasis on friendliness and fun. These two different services will demand the use of varying rhythm and intonation to convey these different kinds of unwritten information.

Accents

The accent of a speaker can also convey information that is not available on the printed page. For example, it can influence perceived personality (Giles and Powesland, 1975). However, different accents have varying effects on different people. It is not until this information is available that a proper evaluation of a service using speech output can be made.

Given all the differences between the spoken and the written word (Table 18.1), evaluating the messages in a dialogue by looking at them on paper can cause problems. Memory demands are not taken into account and the effects of prosody are not appreciated. Therefore, the existence of a prototyper that enables paper based dialogues to be rapidly integrated into a speech-based service with spoken dialogues is paramount for the effective evaluation of dialogues prior to implementation (Benimoff *et al.*, 1989).

Table 18.1 Differences between spoken and written word

Spoken Word	Written Word
Prosodic information available – pitch duration, and intensity.	Difficult to provide prosodic information.
Serial input.	Information can be scanned.
Can have different accents.	Difficult to provide information on accents.
High demand on short term memory.	Little demand on short term memory.

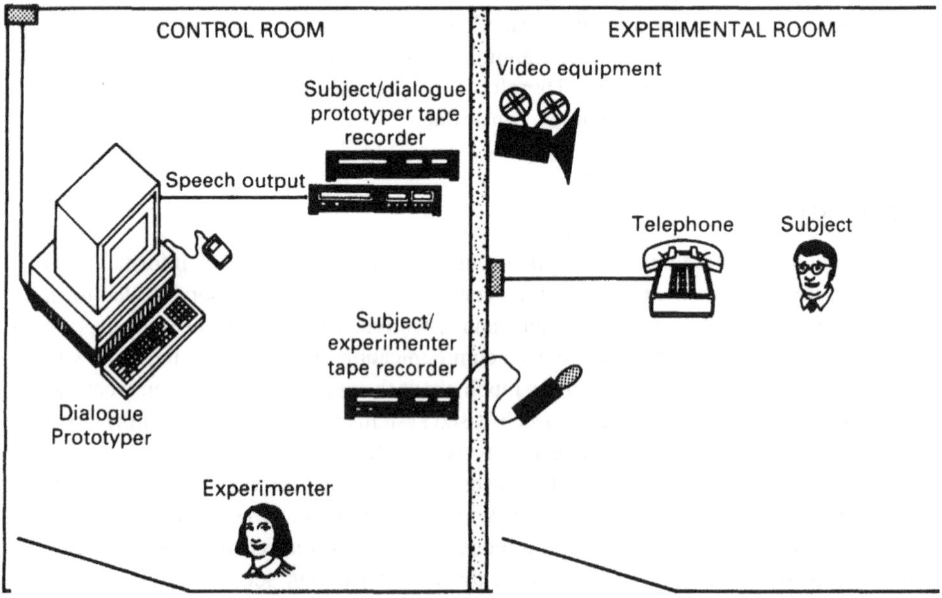

Figure 18.3 A typical usability laboratory.

In short, the translation of the written word or phrase into the spoken phrase can be done in a number of different ways: with different speakers, accents, intonations, pauses, and so on. This means that the evaluation of dialogues on paper can give a false impression to the evaluator. It is only when dialogues are translated into spoken form that a designer can appreciate their effect.

Objective and subjective evaluation

Although it is useful for the designer to assess a prototype system it is important to evaluate the system on the people for whom it is intended; using both objective and subjective methods. At BT, extensive use is made of a subject panel, made up of BT customers, who can be chosen against the demographic characteristics of the intended users of the system being evaluated. Systems are evaluated in a usability laboratory where subjects' interactions are monitored and recorded.

Specifically, the systems are assessed for effectiveness, efficiency, and satisfaction (Guidance on Usability Specification and Measures, May 1992). That is, are the systems effective in actually allowing users to complete the relevant task; are they efficient in allowing them to do it, in terms of the effort incurred by the users; and do the users actually like the system?

Experimental procedures involve:

- audio and video recording of interactions – this provides qualitative information on user interactions;
- speed and error rates of performance as measures of efficiency and effectiveness; and
- questionnaire and interview techniques as measures of subjective satisfaction.

Specifically, evaluations concern aspects such as:

- type of voice – male/female, formal/informal;
- input mode – DTMF vs. Voice for different types of input;
- use of words in messages;
- use of prompts; and
- help messages.

The Speech System Evaluation Methodology (SSEM) (Sidhu, 1991) is employed as a means of subjective assessment. This is based on a methodology developed by Poulson (Poulson, 1987) and consists of three components: questionnaire, interview, and analysis.

The rating scale questionnaire is constructed on the basis of dimensions that are important for the success of speech systems. For example, dimensions such as ease of learning, concentration required, and frustration, are important aspects that must be considered in the evaluation of a speech system. These dimensions form the basis of a list of questions that are given to the users of the system, who respond on a five-point scale.

The responses of the users to the questionnaire are rated and yield numerical values which are summed to form the user's profile. This profile then forms the basis for the interview stage of the SSEM where the interviewer asks questions on extreme responses to the questionnaire.

The methodology can be tailored to address particular aspects of different speech systems but the underlying principle of questionnaire design and subsequent analysis remains static. This subjective methodology is important in providing information on user attitudes to speech-based systems as opposed to information on actual behaviour with a system. However, it is important to address both. Users may find a system to be very usable, but may not actually want to use it.

The results of user trials allow dialogues to be modified and tested again in an iterative fashion.

Step-by-step walkthrough of the evaluation process

The following example, demonstrates how the above methodology has been put into practice, by Human Factor's consultants at BT Laboratories, when developing an automated speech-based system.

- A paper-based dialogue is designed. This is then reviewed by all parties who have a stake in the system, for example, the designer, the customer (the party who commissioned the system) etc. This review is intended only to determine whether the dialogue conforms to the requirements of the customer.
- Once the dialogue has been approved by all the interested parties a prototype of the system is built. The dc is used to develop and run this.
- The experimental design and procedure for the first set of user trials is designed. This involves determining what information is required and how this will be obtained, how many participants are required etc. The briefing instructions, scenarios, questionnaire and interview format are then designed, using the SSEM (Guidance on Usability Specification and measures, May 1992).
- A pilot study is run. The aim of this is to test the prototype for any bugs, and to trial the experimental design. This pilot typically uses members of BT Laboratories.
- The user trials take place. Each participant in the trials is initially briefed by the experimenter. They then interact with the system. This interaction is both audio and video taped for later analysis. A questionnaire is administered, and then the participant is interviewed. Finally the participant is debriefed.
- The log files created by the dc and the subjective data from the questionnaires and interviews are analysed. A number of recommendations for modifications are made on the basis of the findings.
- The recommendations are formally reviewed by the stakeholders. If necessary, modifications to the dialogue structure and messages are made, and then the software for the actual application is implemented.
- After the software has been implemented a further set of user trials takes place. These trials involve participants interacting with the actual application. However,

these trials usually take the same form as the previous trials, in other words, the same experimental design procedure is used. The results of these trials are examined and compared to the previous trial results.

- A decision is made as to whether further modifications and evaluations are required.

Conclusion

This chapter has discussed the dc and described its functions. It is a powerful tool for evaluation, enabling a true representation of dialogues using TTS or recorded speech output. The dc is flexible in providing different configurations of I/O for the construction of a variety of speech systems. This allows for a realistic evaluation of these systems on end-users. In addition, the automated data collection feature of the dc enhances its ability as an evaluation tool.

The evaluations carried out using the dc involve a comprehensive approach addressing objective and subjective measures, with a representative sample of end-users provided by the BT subject panel. The complete package of prototyper and evaluation measures allows for the iterative design of speech systems and means that systems designed at BT are subjected to a very rigorous and realistic assessment prior to implementation.

References

Benimoff, N. I. and Whitten, W. B., II, 1989, Human factors approaches to prototyping and evaluating user interfaces, *AT & T Technical Journal*, Sept/Oct 1989.

Giles, H. and Powesland, P. F., 1975, *Speech Style and Social Evaluation*, Academic Press, London.

ISO/TC159/SC4/N191 Part 11: Guidance on Usability Specification and Measures, May 1992.

Popay, P. *et al.*, 1991, *Voice Applications Style Guide*, BT Internal Report.

Poulson, D. F., 1987, Towards simple indices of the perceived quality of software interfaces, *IEE Colloquium Digest 1987/88*, Evaluation Techniques for Interactive Systems Design: 1.

Sidhu, C., 1991, Speech system evaluation method, BT Internal Report.

19

Intelligent dialogues in automated telephone services

John C. Foster, Rachael Dutton, Mervyn A. Jack, Stephen Love, Ian A. Nairn, Nathalie Vergeynst and F. W. M. Stentiford

Abstract

This chapter describes work carried out as part of a five-year project to investigate the design, implementation and evaluation of dialogues for automated telephone services.

A new Wizard of Oz scheme is described for experiments where volunteer subjects perform specific tasks from their home or work environment using a simulated automated telephone service. The WoZ experimental workbench is based around a simulation of a speech recognition system with selectable performance characteristics, allowing experiments to be carried out with recognizer performance capabilities extrapolated beyond the current state-of-the-art. The WoZ system also possesses a fully integrated dialogue component which can be independently modified to suit differing experimental designs.

A typical experiment using the WoZ scheme is described together with details of the user evaluation measures employed in the research and some preliminary results.

Introduction

This paper describes work carried out as part of a five year research project to investigate the design, implementation and evaluation of dialogues for automated telephone services.

The project involves a series of large-scale field experiments using a new Wizard of Oz (WoZ) scheme – referred to by the acronym JIM (Just sIMulation) – for the investigation of users' attitudes towards simulated automated telephone services and the evaluation of the perceived usability of such services.

Most of the WoZ studies of human–machine speech interaction reported in the literature to date have focused on the problem of characterizing dialogue features in application domains such as rail and air timetable enquiry services (Fraser and Gilbert, 1991). While it is certainly an important aim of the present project to further our understanding of the dialogue aspects of human–computer speech interaction, the project is distinguished from most previous studies by the choice of highly constrained application domains; by the degree of control the JIM software provides over the experimental variables; by the care being taken to quantitatively measure users' attitudes and perceived usability; and by the large subject samples used in the experiments.

Figure 19.1 shows the major components of the experimental programme. The Speech Interface consists of a software simulation of a speech recognition system.

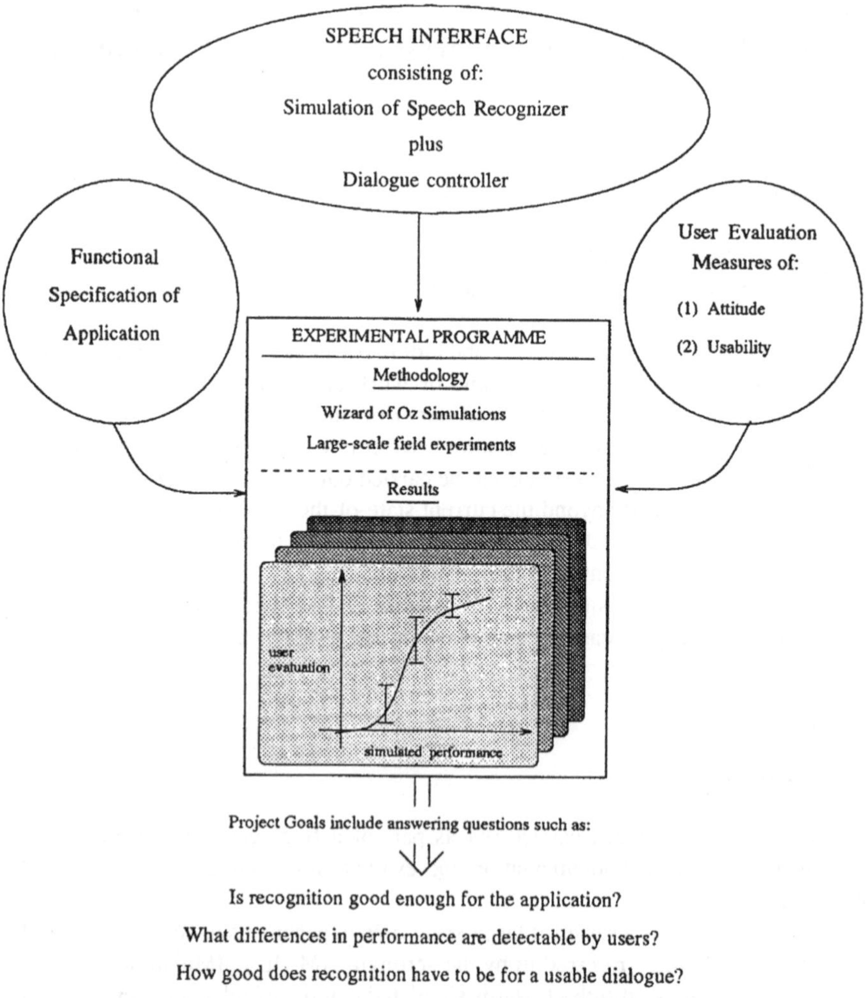

Figure 19.1 Major Components of the Project Research Programme.

The results discussed here are for speaker independent isolated word recognition. The recognizer is coupled to a Dialogue Controller which is specified as a finite-state network and which is responsible for delilvering appropriate pre-recorded prompts to users. Generic applications being considered for automated telephone services with such interfaces include automatic recognition of fixed length numbers with block segmentation (such as credit card numbers), variable length numbers also with block segmentation (such as telephone numbers) and alphanumeric sequences of variable length. The results reported here are for credit card numbers.

The experimental programme itself consists of a series of WoZ experiments in which the characteristics of the speech interface are manipulated and the dependent variables of user attitude and perceived usability are measured by both a telephone questionnaire and a postal questionnaire.

Modification of the speech interface involves changing the characteristics of the recognition system and/or changing the dialogue. Of particular interest in this work is the investigation of users' attitudes to speech interfaces using recognition system capabilities which are beyond the current state-of-the-art. One of the major new features of the project is that it is based on a realistic simulation of an available speech recognition technology allowing experimentation with recognition performance levels extrapolated beyond those currently achievable. We can thereby address, among other key issues, the shape of the usability function for automated telephone interfaces for different levels of recognition performance.

Modifying the speech interface can also involve changing the dialogue used in the telephone service. This allows important user interface and human factors issues to be addressed such as the impact of voice quality on users' perception of the telephone service; the degree to which conventional 'beep' prompts (used in addition to spoken prompts) influence the progress of the dialogue; and the impact of dialogue structure and the specific wording of prompts on users' attitude to the service.

As Fig. 19.1 also indicates, the project goals include other key issues in this field such as the levels of recognition accuracy appropriate to different applications; the different levels of recognizer performance to which users are sensitive; and the level of recognition accuracy required for a given dialogue specification to be judged usable.

The Wizard of Oz experimental workbench

The experimental setup

The WoZ experimental configuration which includes the subject, the experimental operator and the JIM software is shown in Fig. 19.2.

The Software Control Module handles all aspects of the simulation including the initiation of telephone contact at the (volunteer) subject's home or workplace; the delivery of dialogue prompts to the subject; the registering of keystrokes from the experimental operator made in response to spoken input from the subject; the on-line generation of recognition errors when these are required; and the recording of all statistical data on keystrokes and timings.

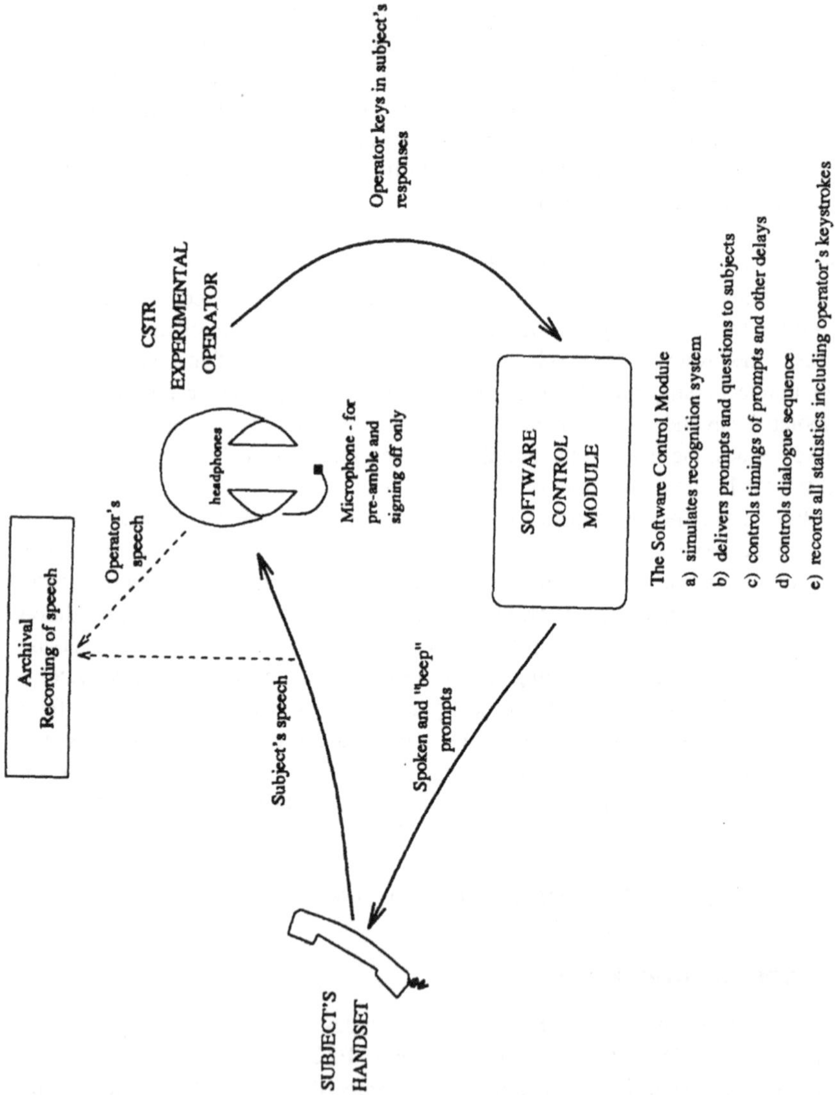

Figure 19.2 The Wizard of Oz experimental workbench.

The JIM software is written in C++ and runs on an IBM-compatible PC with a plug-in board providing connection to the telephone line. In addition to simulating the recognizer, the JIM software controls the archival recording of all interactions between the subject and the operator.

The advantages of this particular WoZ configuration include full control over the experimental conditions; constraints imposed on the operators who key in the subjects' spoken responses; and the fact that all other experimental conditions, such as the introduction of speech recognition errors, are entirely under software control.

An important distinction is made between off-line and on-line simulation of recognition errors. As the experiments use a simulated recognizer and finite-state dialogue networks, it is possible to run new dialogue designs fully automatically with simulated user input. This is the off-line mode of operation. In this way, results are obtained which are characteristic of the particular setting of the simulated recognizer and the chosen dialogue under the implicit assumption that the 'user' supplies consistently perfect input. If the dialogue specification is considered as defining a **dialogue space**, the off-line mode selects particular **paths** through that space. These are the paths (i.e. dialogues) used in the actual experiments. In other words, subjects are guided through the dialogue space along characterized dialogue paths referred to as **scripts**.

It occasionally happens that during an experiment, a subject will misread a digit from their credit card. In this case they go off the pre-defined script. In order to minimize the loss of experimental data in these circumstances, the JIM software switches into on-line mode when this happens. The on-line (real-time) simulator is set at the same recognition accuracy level as the off-line simulator which generated the script in the first instance and uses the same dialogue specification. In other words, when subjects go off script, they are allowed to follow a different path through the dialogue space but the path is valid and can be characterized for that setting of the recognizer and that dialogue. After recognizing a digit in this on-line mode, the JIM software returns to the predefined script for the following digits.

A typical experimental procedure

Each experiment proceeds in three phases. In phase one (the preamble), the operator is connected by the JIM software for normal two-way telephone conversation with the volunteer subject. During this phase the operator checks that the subject is ready to carry out the experiment and has the necessary information to hand, for example, the previously posted replica credit card and the postal questionnaire.

Phase two is the simulated automated telephone service. During this part of the experiment, the operator's microphone is disconnected leaving only the headphones active. This means that the operator can key in the subject's responses to the dialogue prompts but any possibility of the operator leading the subject's responses is removed.

In phase three (the signing off), the operator's microphone is reconnected. During this phase the operator delivers the telephone questionnaire (discussed below) and thanks the subject for participating in the experiment. The subject is reminded to fill in the postal questionnaire as soon as possible after termination of the telephone call.

In addition to selecting scripts from the off-line mode of operation of the Speech Interface, preparations for each experiment involve defining the operator keystroke protocols and specifying the wording of the preamble, the signing off and the telephone questionnaire. Each of these can influence the outcome of the experiment and in particular subjects' responses to the postal questionnaire. Two examples will illustrate these points.

Operator keystroke protocols are the rules used by the operators when entering subjects' spoken responses to the dialogue prompts. Different protocols cause the experiment to follow very different paths through the dialogue space. In experiments carried out to date involving the simulation of an isolated word recognition system, the protocol is defined such that if the expected answer is spoken by the subject in response to a dialogue prompt, even in the context of extraneous linguistic or paralinguistic expression, that expected answer is keyed in by the operator. For example, if a subject in response to the prompt *Please say the next digit* says *It's three* or *Um...er...three* instead of simply *three* as expected, the operator keys in the digit '3'. This protocol implies that we are currently taking a generous view of the capabilities of an isolated word recognition system in the speech interface, including aspects relevant to wordspotting methods. It does mean, however, that fewer subjects go off the script, an important consideration in the early stages of the research work.

A stricter keystroke protocol would require that only 'clean' input be accepted by the WoZ operators. In this case, more subjects would go off script and their data relating to evaluation of the interface would be less relevant to the experimental statistics. Later experiments will compare the effects of these two protocols.

When the JIM software returns control from the automated telephone service to the experimental operator, it is important to decide exactly what to tell subjects about the success or failure of the interaction and about the purpose of the experiment. Experience has shown that the wording at this point in the experiment has a major impact upon users' responses to the usability of the service. For example, in experiments to date subjects have been told that when such services become generally available, their credit card account would be automatically debited. This gives them a wider appreciation of the implications of using this service which is reflected in their response to the postal questionnaire.

Evaluation of attitudes and usability

The two principal instruments for measuring users' attitudes towards the automated telephone service are a telephone questionnaire and a postal questionnaire.

The **telephone questionnaire** consists of a number of questions asked by the human operator immediately after completion of the experiment with the automated service. Among other questions, subjects are asked how many mistakes they thought they made and how many they thought the service made; how long they thought it took them to read their credit card number; and, how they evaluate the service overall. For most of these questions the actual data (such as the number of mistakes and the duration of the call) are already known from the WoZ software. Discrepancies between the data and subjects' perceptions are valuable in evaluating subjects' perception of the effectiveness and efficiency of the service, both of which contribute to measurement of the usability of the service (ISO 1990). Subjects are asked to complete the **postal questionnaire** as soon as possible after using the service. The postal questionnaire consists of several sections the most detailed of which is a set of 22 attitude proposals deriving from Poulson (1987) but additionally containing a number of statements specifically relevant to the evaluation of speech interfaces. A seven-point Likert response scale is provided for each question. An example is the following.

The automated telphone service was easy to use

☐	☐	☐	☐	☐	☐	☐
strongly agree	agree	slightly agree	neutral	slightly disagree	disagree	strongly disagree

The responses to these statements on the questionnaire are used to derive detailed statistical profiles of users' attitudes and the perceived usability of the automated telephone service.

Data analysis and results

The data presented here relate to one experiment involving 256 subjects sampled from the UK population. In this experiment four levels of word recognition accuracy were compared in 5 per cent steps from 85 to 100 per cent using a matched subject set design (matched for sex and age group). The following classes of data were collected.

- Responses to the Likert questions on the usability of the service and subjects' attitudes across the four settings of the recognizer.
- A rank ordering of attributes perceived by subjects to be particularly salient for automated telephone services.
- Responses to the telephone questionnaire on which a number of correlational analyses have been based.
- A breakdown of the dialogue events and features including the effects of the subjects' environment on their ability to respond accurately to the dialogue prompts, the various vocabulary items used to indicate 'zero' as a digit and problems related to the use of 'beep' prompts.
- Transcriptions of relevant parts of the audio recordings for dialogue and usability analysis.

As an example of the descriptive statistical analysis currently being carried out on these data, Fig. 19.3 shows part of a usability profile. Mean attitude scores for five attributes of the automated service and for two accuracy levels of the simulated recognizer are compared for the same dialogue.

Figure 19.3 suggests that subjects found the version of the service with no queries or rejections of digits (i.e. simulated word recognition accuracy 100 per cent) to be easier to use than the version with several queries (simulated word recognition accuracy 90 per cent). Subjects also perceived the former to be more reliable. Subjects were neutral when asked about their preference for talking to a human operator – for both settings of the recognizer. The reactions to the voice used for delivering the prompts and perceptions about the helpfulness of the beep prompts are the reverse of the other evaluations: the service at 90 per cent word recognition accuracy received a more positive evaluation than the service operating at 100 per cent word recognition accuracy. One explanation for this is that at the higher accuracy level, subjects require less concentration since they are not being queried about the digits they are reading, they therefore have no reason to develop any positive attitude towards the voice and the beep prompts. Conversely, subjects at the lower accuracy level find the beep

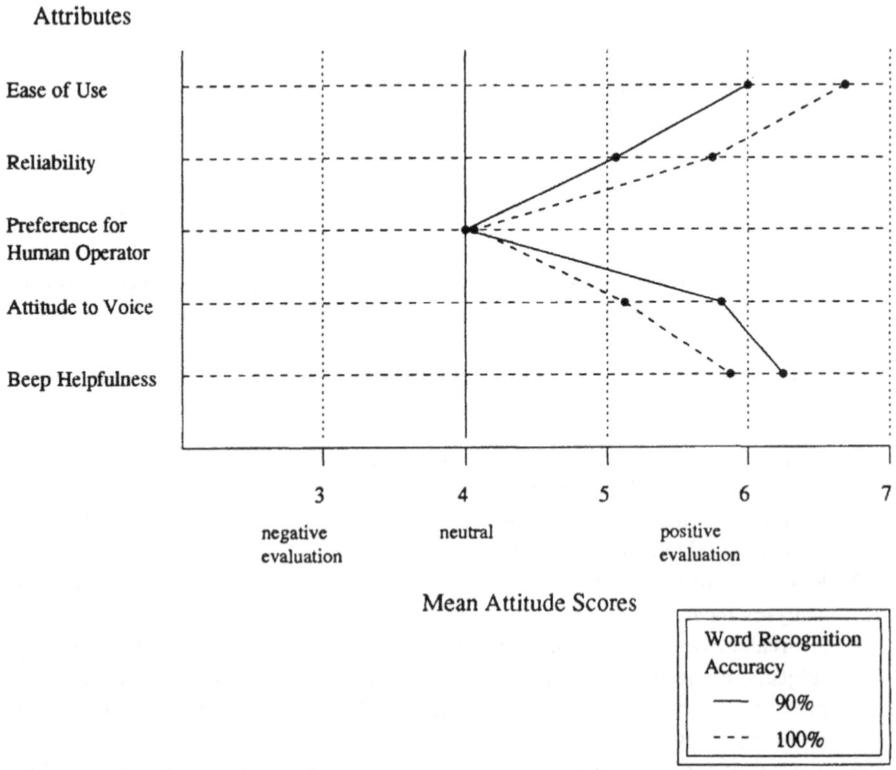

Figure 19.3 A Usability Profile for the Automated Telephone Service.

prompt and the voice of positive assistance in their navigation of the problems posed by the queries and rejections of their input; they therefore have a more positive attitude to these attributes of the telephone service.

Conclusions

Preliminary results from the new WoZ scheme described in this paper using Likert response scales and usability profile methods have demonstrated the applicability of telephone and postal questionnaires in evaluating the usability of automated telephone services.

Acknowledgements

The authors wish to acknowledge the support for this research from BT's Strategic University Initiative and the contributions made by other members of the Dialogues for Systems Team at BT Laboratories (Martlesham).

References

Fraser, N. M. and Gilbert, G. N., 1991, Simulating speech systems, *Computer Speech and Language*, **5**, 81–99.

ISO, 1990, Ergonomic Requirements for Office and Visual Display Terminals (VDTs), Part 11: Usability Statements, *International Standards Organization*, ISO CD 9241-11 version 2.5 8 July 1990.

Poulson, D., 1987, Towards simple indices of the perceived quality of software interfaces, in *IEE Colloquium – Evaluation Techniques for Interactive System Design*, IEE, Savoy Place, London.

20

Design guidelines for using speech in interactive multimedia systems

Kevin Hapeshi

Abstract

The combination of speech, sound and hypertext software turns a personal computer into an interactive multimedia system providing a medium for presenting instructional material in different modalities and styles. Verbal material can be presented in written or spoken form; in many cases the choice can be determined by the system developer or by the users themselves. The flexibility and interactive nature of multimedia systems suggest that the simple transfer of techniques developed in more traditional audio-visual media, or even the more recent interactive video, would be inappropriate. A good example is the potential offered by speech input, whether in the form of speech recogniton or digitized recording. While in systems currently available, speech input is rarely used, cognitive research points to many ways in which the effectiveness of instructional material can be significantly enhanced by its addition.

In any case, what is needed is a clear set of guidelines for both designers and users, which will enable the most effective use of this new medium. This chapter provides a number of guidelines on the basis of a review of experimental studies on mixed media verbal presentation and its effects on the efficiency of learning and memory.

Introduction

Multimedia systems are interactive computer-based development and delivery platforms that combine a rich mix of text, graphics, stationary and moving pictures, speech, music, and sounds. While the use of graphics, text and sound has been commonly available on personal computers, it is the addition of broad-band audio and video channels that characterize multimedia systems. The present disussion is concerned mainly with the way broad-band audio channels can be utilized as a major information channel.

There are essentially three ways of presenting high quality audio in multimedia systems:

1. through the analogue sound channel of attached video devices;
2. through compact disc digital audio technology (CD–DA) and its derivatives;
3. through on-board or plug-in broad-band analogue to digital (A/D) and digital to analogue (D/A) devices (hereafter referred to as a sound interface).

Only the latter provides the flexibility for speech input as well as output. Sounds and speech can be digitized into files, which can be edited, manipulated and played back. The auditory channel can be presented either in isolation or in conjunction with text or graphics.

The sound interface is by far the most flexible system and is currently enjoying a boom in the personal computer market. Typically, such devices must be capable of digitizing at a rate of 22 KHz in two channels. For these devices to be effective, they must also provide direct memory access (DMA), facilitating the independent retrieval of digitized data from memory, thus freeing the central processor to perform activities such as modifying the visual display.

The only major limitation is the size and access speed of the device holding the digitized sound file. With today's technology, for example, access to information held on hard discs is fast enough, while floppy discs are too slow. Files can be loaded into random access memory (RAM) prior to replay; however, the size of available memory may itself be limited. Perhaps the most important limitation on the use of long digitized files is the size of storage space required. Some 60 Mbytes of hard disc storage is needed for one hour of speech, which has been digitized at a rate of 16 KHz. Digital compression techniques can help by reducing storage needs by around 50 per cent. If extensive speech output is required speech synthesis can be used, which theoretically provides an unlimited vocabulary for very little storage space (perhaps as little as 2–5 Kbytes). Speech synthesis enables text files to be presented in spoken form, and at a rate of 120 words per minute, a 10 000 word text file (about 65 000 bytes) would represent around 1.5 hours of spoken material.

In general, interactive multimedia systems offer additional flexibility over other broad-band systems. For example, in interactive video, audio and visual data are stored together so cannot easily be separated. While in multimedia systems, audio and visual channels are separate so sound can be played independently of the visual images, although synchronization is also possible. An advantage of this is that as an information channel, the auditory channel can be used while the listener is performing other activities, even away from the delivery machine. This enables verbal instruction to be given, while the listener performs a task, though this is only practicable if the information in the auditory channel is designed to 'stand-alone', relatively independent of any visual information.

The use of auditory interfaces in computer systems has been attracting more attention (Buxton *et al.*, 1985; Gaver, 1986). Professor Dylan Jones refered to this as the 'sonic interface' (Jones, 1989) and argues that it tends to be over-sold. The characteristics of human audition lead to the conclusion that the use of the auditory channel requires careful consideration of the limits in human information processing and the application of human factors guidelines (Hapeshi and Jones, 1988). Despite some reservations on the usability of the auditory channel in computer interfaces, there are some advantages which can be usefully employed in learning systems. Unlike vision, audition does not rely on the physical orientation of the recipient to the source so it does not matter from which direction the sound emanates. Therefore the auditory channel is ideal for transmitting information to a listener performing some activity which occupies the visual sense. For this reason, 'distance learning' materials as well as novels supplied on cassette tapes are very popular. It is likely that interactive multimedia will increasingly be used for the delivery of distance learning packages and generally this implies mostly speech output. However, studies in human cognitive processes suggest that speech input can also be used to good effect in learning and training systems.

What is needed is a number of guidelines to help developers maximize the effectiveness and usability of systems which incorporate speech input and output, and to suggest ways these can be combined with broad-band visual channels. The present discussion is confined to systems intended for training and instruction, since it is in these systems that efficient design is most important if the goal of long-term education is not to be compromised. In effect, computer-based training systems have replaced the programmed learning machines as the ultimate in self-instruction, and multimedia represents another step in the development of computer-aided instruction.

Guidelines

Use speech and sound sparingly

A fairly obvious point to be made here is that the sound-channel should be used sparingly when the system is in a public place (such as an open-plan office) to avoid sound pollution. Headphones can of course be used, but this is not always practical or desirable. Secondly, privacy must be respected. For example, multimedia training systems should avoid using the sound channel to provide feedback on learning progress, otherwise a user's failure to learn may be obvious to all within hearing range. range.

The extensive use of the speech channel can have negative side-effects in terms of user behaviour and attitudes, particularly where synthesized speech is used. A number of studies have shown that there is a tendency for talking computers to be given human attributes. Despite being obviously computer-generated, synthesized speech is often regarded by some listeners as being 'unfriendly', 'evil and sinister', or 'harsh' (Michaelis and Wiggins, 1982; Edman and Metz, 1983). According to Peacock (1984) some listeners resent speech devices 'telling them what to do'. This 'anthropomorphism' is the result of users endowing the speech device as having greater intelligence and power than it actually has. A problem with this is that as speech output becomes more realistic, inappropriate user attitudes and expectations will be generated, which can in turn lead to inappropriate behaviour to a system (Hapeshi and Jones, 1988).

In terms of general instructional material, the sound channel should not be used unnecessarily. In their suggestions for developers of instructional films, Carpenter and Greenhill (1956) stated that commentary should not 'pack' the sound track. Carpenter and Greenhill pointed to experimental studies showing that increasing the number of words (per minute of film) had a negative effect on learning. There is some more recent support for this from a study reported by Aldrich and Parkin (1987). They found that subjects were able to answer questions about an auditory passage as accurately following (a) hearing of an expert summary, compared to (b) hearing the whole passage alone, or even (c) the passage plus summary. It is clear from this that spoken messages could be minimized to present only the relevant material, rather than presenting longer messages and expecting the listener to extract the relevant details. It appears that short, carefully constructed auditory messages can be just as effective for learning as longer ones, and as such are more efficient. The main advantage of short expert summaries is that they will require less user time, enable more material to be covered, and allow more material to be stored in the system itself. The disadvantage is the additional effort required at the development stage.

It is not just speech that can mar the effectiveness of instructional material, music can also be detrimental. In another guideline presented by Carpenter and Greenhill (1956) they argued that their studies suggest the addition of music yields little benefit in terms of the instructional effectiveness of an informational film. This presumably refers to background music that is irrelevant to the information. The study reported by Hayes, Kelley and Mandel (1986) confirms this; they found that irrelevant sounds detracted from important thematic detail.

Use speech for short pieces of information

A crucial factor in the use of speech output for presenting verbal information is its transitory nature as well as the way the human memory system works. In general, new information can only be held in memory for periods lasting only a few seconds before it begins to deteriorate. It is a well established fact that there is an advantage of auditory over visual presentation for recall during brief retention intervals (Engle, 1974; Crowder, 1970; Murdock and Walker, 1969), which can be attributed to some form of precategorical auditory or 'echoic' memory (Crowder, 1976; Neisser, 1967). The visual equivalent is known as **iconic memory** (Neisser, 1967). Each serve to retain information over very short periods of time even if the information was not attended at the moment of presentation, but the echoic representation for speech persists longer than iconic representation for text (Wickens, 1992; Neisser, 1967). This implies that short lists are best presented in auditory form, longer ones in visual form (when reviewing is possible).

During short-term storage, information can be assimilated into longer-term memory and thus will be more enduring, but this process may require some effort on the part of the listener, such as active rehearsal (Waugh and Norman, 1965; Atkinson and Shiffrin, 1968). If new verbal information is in some way 'prominent' or meaningful to the recipient, long-term 'incidental' learning may occur without active effort. However, where information is unfamiliar, some form of phonological storage may be essential for long-term encoding (Baddeley *et al.*, 1988). Presenting important information in speech form will facilitate this.

Provide reviewing facilities for the auditory channel

Speech presentation is serial, which means that messages are transient–distributed over time. This places a burden on memory that visual presentation does not pose. For example, Kidd (1982) found that listeners had difficulty understanding and remembering spoken menus if they consisted of unclear, overlapping categories, while clear non-overlapping categories posed fewer problems. Synthesized is less intelligible and makes the information processing burden even greater (Luce *et al.*, 1983). It may be that in carefully designed multimedia presentation, visual stimuli can provide a context which aids the intelligibility and processing of normally difficult spoken material. This possibility deserves greater attention from researchers.

In general, limits in human information processing apply to verbal information presented both aurally and visually, but because visual information is usually available for inspection over longer periods, the learning has a greater opportunity to commit the material to memory. This was demonstrated in a study by McDowd and Botwinick (1984). They looked at the ability of subjects to recall prose presented as text or in spoken form. Results showed that visual displays permitting review (i.e., reading) were better for recall than visual displays or auditory input permitting less or no review. Clearly, the advantage of visual displays for longer messages disappears if review is not possible. This fact is partly supported by Das and Siu (1989) who used children as subjects. They reported no difference in the ability of good readers to recall detailed information from a visual passage, read only once, and an auditory passage heard once.

It is reasonable to assume that simple facilities for reviewing in the auditory channel could enhance long-term retention of speech material, but it is unclear whether this could rival text displays. An alternative means for improving learning from auditory displays is to take advantage of the fact that physical features of auditory messages (such as sex of the speaker) appear to be encoded with the linguistic content itself. For example, Geiselman and Crawley (1983) have shown that the physical characteristics of the voice are stored with auditory verbal material, so that recognition is better if the target stimulus is replaced in the same voice in which it was originally presented.

Presentation modality itself may be an important factor in recall since it too seems to be an important part of the coding of learned material. Dean *et al.* (1988) used a recognition task, and found that subjects performed better if the test stimulus was presented in the same mode as originally seen or heard. They concluded that the physical characteristics of a stimulus persist in memory beyond short-term intervals. Similar conclusions were drawn by Lehman *et al.* (1985). These results imply that the particular mode of presentation may be used as an aid to learning and recall, since it can be used as a mnemonic 'handle' by which information can be retrieved.

Where possible, any detailed prose to be presented should be in the form of text which can be reviewed. Short summaries of this can be presented in auditory form (see p. 179). However, if this is not possible, reviewing facilities should be provided for the auditory channel.

Provide mixed mode presentation where it can enhance processing

Using film and video there are four basic forms in which speech is combined with images:

1. a monologue or lecture with the speaker shown in the display;
2. a dialogue or conversation between those on screen;
3. a descriptive narrative, which explains directly events on the screen; and
4. speech presentation, with related illustrations presented on screen.

In the first two forms, a visual display will facilitate processing of the auditory message if the speaker's face can be seen, because facial expressions, particularly lip-movements, enhance speech intelligibility (Binnie, 1974; Ritsma, 1974). Visual displays of text in the auditory message can also enhance speech intelligibility. A common experience is being able to recognize the lyrics of a song more easily if the listener can simultaneously examine the printed words. All of this suggests that when the visual and auditory channels provide congruent verbal messages, processing is easier.

There is also evidence that the presence of moving images can serve to enhance comprehsnsion and learning of spoken material, particularly in children. Hayes *et al.* (1986) compared the effectiveness of television to that of radio presentations of narrative information. Generally, the inclusion (during recall) of inaccurate story content and the distortion of actual story detail occurred more often in the auditory only condition than in the aural and visual condition. Hayes *et al.* reported that subjects in the radio condition recalled dialogue and sound effects in greater detail than in the television condition, but they judged that the dialogue and auditory details recalled tended to convey little of the overall theme. This suggests that simultaneous visual events can help listeners to select only appropriate auditory events for processing and later recall, and that 'irrelevant' sound effects can detract from more relevant details.

Support for the value of integrated visual and auditory information has been provided in other studies using different paradigms. Schwartz and Kulhavy (1987) found that aurally presented story content was better recalled if subjects viewed a visual map, which was congruent with the story content, particularly if the narrative structure was relatively simple. Schwartz and Kulhavy concluded that learners consistently used the spatially-based information as a way of increasing recall of the passage material. Hoffner *et al.* (1989) demonstrated that if visual displays did not support auditory displays, learning was impaired. They reported that recognition memory for auditory material was significantly reduced by the presence of conflicting video presentation.

It is not clear whether there are benefits in mixed presentation for the last two forms for combining speech and vision in film. If the visual and auditory material are competing for limited capacity human information processing resources, such mixed-modal presentations may be less effective for long-term learning than single-mode equivalents. Where vision and audition compete, the effect of a descriptive narrative may be similar to the effect of irrelevant speech. For example if there is a tendency for users to attempt to recode verbally the visual events, this information can be lost in the presence of background speech (Salamé and Baddeley, 1982; Jones *et al.*, 1989). In such cases, simultaneous auditory descriptions should be kept to a minimum (see below).

Multimedia systems provide an opportunity to combine the relative advantages of visual and auditory presentation in ways that can lead to enhanced learning and recall. For example, it has been shown that visual and auditory presentation can be usefully mixed to counter forgetting due to interference. In a study reported by Dean *et al.* (1983) it was shown that presentation of information either visually or aurally, in a random fashion, can significantly reduce interference effects. In their first experiment Dean *et al.* (1983) presented undergraduates with a 40-word list presented in the auditory and the visual modality. In the 'blocked' presentation group, subjects were presented half the words in one modality followed by the remaining 20 words in the other modality. For the 'random' group subjects also received 20 nouns in each modality, but the presentation was random. Following a delay, all subjects were required to complete a recognition test. Analysis showed a distinct reduction in serial position effects when the modality of presentation was random. In contrast, the blocked presentation produced well-defined normal serial position curves.

In a second experiment Dean *et al.* (1983) studied the effects of a shift in the modality of presentation on proactive interference. Four similar prose passages were presented with a modality shift taking place in the last passage for the shift condition. Subjects in nonshift conditions were presented the passages in either the auditory or visual mode. Results show that a shift in the modality of presentation of a prose passage provided a powerful releaser from proactive interference. Proactive interference is caused by earlier learned items interfering with items learned later. The more similar the early and late items are, the more likely interference will occur. By changing the style and mode of presentation this interference can be reduced. Thus, if a number of lists or facts are to be learned in a session, alternating mode of presentation can improve learning and recall.

Avoid competition between visual and auditory channels

While there is extensive evidence for the benefit of audiovisual presentation, Rollins and Thibadeau (1973) also provide evidence that the audio and visual channels can be used to present different information simultaneously, as long as these do not compete for the same limited processing resources. In their study Rollins and Thibadeau (1973) asked subjects to repeat (shadow) an auditory prose message heard in one ear while receiving a second stream of messages. This second stream could be:

1. an auditory list of concrete nouns in the second ear;
2. printed words;

3. pictures of objects easily labelled (verbally recoded); or
4. pictures of objects difficult to label.

It was found that the shadowing task interfered with recognition scores for the first three groups, but had little effect on the performance of the last. Furthermore, shadowing interfered more with information received aurally than with any form of visual presentation. They concluded that information received visually is encoded in a long-term modality specific form that operates independently of the auditory mode. Interference from auditory shadowing is only evident when subjects attempt to recode verbally the visual items.

This result of Rollins and Thibadeau (1973) can be interpreted within the framework of the working memory model (Baddeley, 1986; Hitch, 1980). Within this model, both auditory shadowing, and verbal recoding compete for the limited processing capacity of the 'central executive' and more importantly the 'articulatory loop'. Thus when shadowing, subjects performing the first three tasks were less able to encode the secondary stream for long-term retrieval. Subjects were able to perform the fourth task, however, since the 'natural' strategy of verbal recoding was inappropriate, so some form of visual coding must have been used instead. This form of coding does not require the 'articulatory loop', therefore was not affected by shadowing.

The implications of these findings for the design of multimedia presentation is that simultaneous speech and visual presentation should be avoided if the visual event or display must itself be interpreted in verbal form (verbal recoding). In other words visual images can be presented with auditory messages without detriment to learning and recall, as long as the visual display does not demand or encourage verbal recoding.

Redundant or irrelevant auditory information should be avoided

Sounds, whether speech, music or special effects, can be distracting if not relevant to the instructional theme because audition in the human brain is intimately connected to the arousal and activation systems. That the auditory channel is constantly monitoring events means it can be regarded as the 'sentinel of the senses', and is therefore especially suited to attention-grabbing warning signals (Jones, 1989).

A number of studies have demonstrated that these attention-grabbing qualities of the auditory channel combined with its special adaptation to speech means that it is highly disruptive to information presented in the auditory modality when listeners are performing an activity which requires verbal skills. Salamé and Baddeley (1982) reported that unattended background speech impairs performance in a short-term serial recall task, while the same level of 'pink' noise had little effect. More recently Jones *et al.* (1989) showed that background speech impairs reading as well as memory and, significantly, this did not depend on the sound level. Even speech equivalent to a loud whisper, will disrupt reading performance.

Studies have also demonstrated the disrupting effect of speech output during data-entry from short-term memory using an automatic speech recognizer (Jones and

Hapeshi, 1989). Item-by-item speech feedback, which echoes the correct input, is significantly more disruptive to recall than an immediate tone feedback.

Use speech input for active participation

In most interactive multimedia systems, speech input refers to digitizing speech messages for later playback. However, in some systems automatic speech recognition (ASR) is available, so speech input can be used to control the system or as an alternative input device. Today, ASR as a means of input to a computer is a serious prospect and many sound interfaces that are currently available include speech recognition facilities. Speech input is often regarded as a potential replacement for manual input devices such as the keyboard, or the mouse. There are two main reasons for this. Firstly, speech input can be used when other means of input are impractical, such as when the operator's hands and eyes are busy performing and monitoring other tasks. Secondly, with very few exceptions, speech is a ubiquitous communication skill in human operators, so theoretically requires little additional training. However, there are a number of serious limitations in currnt technology, which can make the use of ASR either impractical, or difficult (Hapeshi and Jones, 1988).

As speech is a natural communication medium between humans, it is often assumed that it would also be a natural communication form in human–computer interactions. Certainly science fiction writers have used it to great effect, but as yet speech interfaces represent a major human factors design problem. Generally, speech systems are sensitive to more sources of variability than other input systems, and where problems do occur, recovery can be very problematic. Many of these problems can be overcome by the implementation of good human-factors principles (Hapeshi and Jones, 1988; Jones, Hapeshi and Frankish, 1989), but it is unlikely that ASR will replace the need for other means of input to computer systems.

Within multimedia, it seems likely that the use of ASR will reflect the success of ASR in general. A number of surveys have identified successful ASR applications to be small-vocabulary, isolated-word applications common in the industrial setting. Generally, these applications are characterized by the fact that there already exists a highly constrained protocol (Frankish, 1987; Noyes and Frankish, 1989).

In contrast, the office environment appears to be a relatively hostile one for ASR (Noyes and Franish, 1989). In the office speech is used in several ways, which makes utterance consistency (necessary for speaker-dependent systems) very difficult. One possible exception is applications which require only a small-vocabulary, isolated-word for commands or data-entry. For example, computer-aided design. However, even this may be impractical in a busy, noisy office, and for use over prolonged periods (Newell, 1984; Hapeshi and Jones, 1988).

How could ASR be used in multimedia systems? One likely use will be in the invocation of frequently used command or menu selections. Another is in the simulation of activities involving spoken requests or commands. A third use will be in language education and training, particularly in reading and foreign language teaching. In terms of general learning, facilities for speech input can be used to encourage users to vocalize material to be learned, a procedure known to be very effective for long-term retention (Gathercole and Conway, 1988; Landauer and Bjork, 1978). Thus requesting active participation from the user by encouraging vocalization of material to be learned will enhance recall. Of course, speech input can also be encouraged where only speech digitization is available, but ASR provides a means for automatically monitoring utterance consistency and accuracy, as well as a medium for constant on-line testing.

Auditory messages should be independent

A major benefit of speech in education is that it can be used to teach the non-literate (Saettler, 1968). As a training tool audio-only delivery can be as effective as delivery via TV and film. Given certain (non-verbal) activities, it is feasible to use speech to present instructional material, while the operator is working on other things. For example, certain education packages (such as Open University units) are available in audio cassette form, and the students are encouraged to use moments - such as driving the car or doing housework - as valuable learning periods. Multimedia provides the same facilities since it can be used to present sound only; speech need not be presented in combination with pictures or text. Thus the designer should not assume the visual and audio channels will always be presented together.

Clearly, this is only practical if the information in the auditory channel is designed to 'stand-alone', relatively independent of any visual information. This contradicts a suggestion made by Romiszowski (1986), who advises the use of 'natural' commentary in instructional video, eliminating the need for linking phraseology. For example, the use of scripts such as 'pick up *this* and hold it like *this*' while a finger points to an object and a hand picks it up. In film and video, sound and visual images always go together. In some multimedia systems they need not, so sound must be more informative.

Conclusion

Interactive multimedia broadens the bandwidth for presenting information to the user. It makes available a mix of text, graphics, full-motion video and sound and therefore provides a rich medium for training and instruction, and its use may be dominated by existing instructional film, audio-visual, and interactive video formats. However, with the existence of relatively cheap 'desk-top video' software, it can be argued that it is the addition of sound and speech facilities that turns a personal computer system into an inexpensive multimedia platform. Therefore, we can expect, within the very near future, a large proliferation of multimedia publications.

The general aim of this chapter was to highlight cognitive factors and experimental evidence relating to cross-modal information presentation and to consider ways in which visual and auditory channels can be used to deliver instructional messages in multimedia systems. It is argued that the flexibility of multimedia systems requires a new approach. The synchronous audition and vision used in traditional instructional film or even interactive video may be inappropriate for multimedia. A number of guidelines for the use of the auditory channel in multimedia were constructed on the basis of experimental studies which reflect on the effectiveness of different modes of presentation for learning and retention. These guidelines are intended mainly as an aid to developers of multimedia documents or programmes, but should also be useful to users.

References

Aldrich, F. K. and Parkin, A. J., 1988, Improving the retention of aurally presented information, in M. M. Gruneberg, P. E. Morris and R. N. Sykes (Eds), *Practical aspects of memory II: Current research and issues*, Chichester, John Wiley & Sons.

Atkinson, R. C. and Shiffrin, R. M., 1968, Human memory: a proposed system and its control processes, K. E. Spence and J. T. Spence (Eds), *The Psychology of Learning Motivation*, Academic Press, NY, 92-122.

Baddeley, A. D., 1986, *Working Memory*, Oxford: OUP.

Baddeley, A. D., Pagagno, C. and Vallar, G., 1988, When long-term learning depends on short-term storage, *Journal of Memory and Language*, **27**, 586–95.

Binnie, C. A., 1974, auditory-visual intelligibility of various speech materials presented in three noise backgrounds, *Scandinavian Audiology*, **4**, 255–80.

Buxton, W. *et al.*, 1985, Communicating with sound, *Proceedings of CHI '85*, 115–9.

Carpenter, C. R. and Greenhill, L. P., 1956, *Instructional Film Research Reports*, Technical Report 269-7-61, NAVEXOS P1543, Special Devices Centre, Port Washington, NY.

Crowder, R. G., 1976, *Principles of Learning and Memory*, Hillsdale, NJ: Erlbaum.

Crowder, R. G., 1970, The role of one's own voice in immediate memory, *Cognitive Psychology*, **1**, 157–78.

Dean, R. S., Garabedian, A. A. and Yekovich, R., 1983, The effect of modality shifts on proactive interference in long-term memory, *Contemporary Educational Psychgology*, **8**, 28–45.

Dean, R. S., Yekovich, F. R. and Gray, J. W., 1988, The effect of modality on long-term recognition memory, *Contemporary Educational Psychology*, **13**, 102–15.

Edman, T. R. and Metz, S. V., 1983, A methodology for the evaluation of real-time speech digitization, *Proceedings of the Human Factors Society*, 27th metting, 104–7.

Engle, W., 1974, Modality effect: is precategorical acoustic storage responsible? *Journal of Experimental Psychology*, **102**, 824–9.

Frankish, C., 1987, Voice input applications and human factors criteria, *Proceedings of International Speech Tech '87*, NY: Media dimensions, 133–6.

Gathercole, S. E. and Conway, M. A., 1988, Exploring long-term modality effects: vocalization leads to best retention. *Memory & Condition*, **16**, 110–9.

Gaver, W., 1986, Auditory icons: using sound in computer interfaces, *Human Computer Interaction*, **2**, 167–77.

Geiselman, R. E. and Crawley, J. M., 1983, Incidental processing of speaker characteristics: voice as connotative information, *Journal of Verbal Learning and Verbal Behaviour*, **2**, 15–23.

Hapeshi, K. and Jones, D. M., 1988, The ergonomics of automatic speech recogntion interfaces, in D. J. Oborne (Ed.), *International Reviews of Ergonomics*, **2**, 251–90.

Hayes, D. S., Kelly, S. B. and Mandel, M., 1986, Media differences in children's story synopses: radio and television contrasted, *Journal of Educational Psychology*, **8**, 28–45.

Hitch, G. J., 1980, Developing the concepts of working memory, in G. Claxton (Ed.), *Cognitive Psychology: New Directions*, 154–96, London: Routledge and Keagan Paul.

Hoffner, C., Cantor, J. and Thorson, E., 1989, Children's responses to conflicting auditory and visual features of a televised narrative, *Human Communication Research*, **16**, 256–78.

Jones, D. M., Miles, C. and Page, J., 1989, Disruption of reading by irrelevant speech: effects of memory, arousal or attention? *Journal of Applied Cognitive Psychology*, **4**, 89–108.

Jones, D. M., Hapeshi, K. and Frankish, C., 1989, Human factors and the problems of evaluation in the design of speech recognition interfaces, in D. Diaper & R. Winder (Eds), *People and Computers*, **3**, 41–50, Cambridge: CUP.

Jones, D. M. and Hapeshi, K., 1989, Monitoring speech recognizer feedback during data entry from short-term memory: a working memory analysis, *International Journal of Human Computer Interaction*, **1**, 187–209.

Jones, D. M., 1989, The Sonic Interface, in M. J. Smith & G. Salvendy, *Work with Computers: Organizational, Management, Stress and Health Aspects*, Amsterdam: Elsevier Science, 382–8.

Kidd, A. L., 1982, Problems of man–machine dialogue design, *Proceedings of the 6th International Conference on Computer Communications*, London, 531–536.

Landauer, T. K. and Bjork, R. A., 1978, Optimum rehearsal patterns and name learning, in M. M. Gruneberg, P. E. Morris & R. N. Sykes (Eds), *Practical aspects of memory*, Academic Press, London, 625–32.

Lehman, E. B., Mikesell, J. W. and Doherty, S. C., 1985, Long-term retention of information about presentation modality by children and adults, *Memory & Cognition*, **13**, 21–8.

Luce, P. A., Feustel, T. C. and Pisoni, D. B., 1983, Capacity demands in short-term for synthetic and natural speech, *Human Factors*, **25**, 17–32.

McDowd, J. and Botwinick, J., 1984, Rote and gist memory in relation to type of information, sensory mode, and age, *Journal of Genetic Psychology*, **145**(2), 167–78.

Michaelis, P. R. and Wiggins, R. H., 1982, A human factors engineer's introduction to speech synthesizers, in A. Badre & B. Schneiderman (Eds), *Directions in Human–Computer Interaction*, Norwood, NJ: Abex Publishing.

Murdock, B. B. and Walker, K. D., 1969, Modality effects in free recall, *Journal of Verbal Learning and Verbal Behaviour*, **8**, 665.

Neisser, U., 1967, *Cognitive Psychology*, Appleton Century Crofts, NY.

Newell, A. F., 1984, Spech – the natural modality for man–machine interaction, *Proceedings of the IFIP Conference on Computer Interaction*, INTERACT, 4–7 Sept., London, 174–8.

Noyes, J. M. and Franish, C., 1989, A review of speech recognition applications in the office, *Behaviour and Information Technology*, **8**, 475–86.

Peacock, G. E., 1984, Humanizing the man/machine interface, *Speech Technology*, **2**, 106–8.

Ritsma, R. J., 1974, Speech Intelligibility with eye and ear, *Scandinavian Audiology*, **4**, 231–42.

Rollins, H. A. and Thibadeau, R., 1973, The effects of shadowing on recognition of information received visually, *Memory & Cognition*, **1**, 164–8.

Rosmiszowski, A. J., 1986, *Developing Auto-Instructional Materials*, Kogan Page, London.

Saettler, P., 1968, *A History of Instructional Technology*, McGraw-Hill, NY.

Salamé, P. and Baddeley, A. D., 1982, Disruption of short-tterm memory by irrelevant speech: implications for the structure of working memory, *Journal of Verbal Learning and Verbal Behaviour*, **21**, 150–164.

Schwartz, N. and Kulhavy, R. W., 1987, Map structure and the comprehension of prose, *Education & Psychology Research*, **7**, 113–28.

Wickens, C. D., 1992, *Engineering Psychology and Human Performance*, 2nd Edn. Harper Collins, NY.

21

Speech technology in the future

Jan Noyes

Overview

One approach to predicting the future is to look back to the past. Studying the development of a system and/or its component parts may identify patterns and trends in its history, which could be used to make forward projections concerning future use and utilization. In keeping with this approach, this chapter begins by briefly reviewing the development of speech-based systems. It continues by providing an overview of the current state-of-the-art of speech technology, and concludes by attempting to make some forecasts with regard to future developments in interactive speech technology.

The practical problems associated with the production of machines to (i) recognize human speech and (ii) generate speech output present very different problems. One analogy has been to liken the production of speech output to 'squeezing toothpaste out of the tube' in contrast to speech recognition, which is akin to 'pushing the toothpaste back into the tube' (Bristow, 1986). The development of the technology for speech output does not pose as complex a problem as that needed for the recognition of speech, and as a result, speech output techniques are generally more advanced than speech recognition in terms of their usability and refinement. However, comparisons are not always valid due to the very different nature of the problem and the type of application being addressed.

Speech communication

The continuing quest to communicate with machines by speech can be explained in part by the many innate advantages associated with the spoken word. Learning to speak and to listen are skills which the majority of humans master with little effort, and it is with ease that verbal communication is used throughout adult life. Furthermore, spoken language was established long before other modes of communication such as writing and reading. One consequence of this is that human speech is perceived as being both natural and ubiquitous across the human population, and the development of communication with machines by speech is frequently viewed as a natural progression in human–machine interaction. The familiarity and convenience of speech are likely to give it superiority over other information exchange modes, and using speech allows communication with machines in the user's language not the system's, which may be particularly relevant in the case of advanced technologies where control has previously been through the medium of a programming language. The use of a 'natural language' dialogue will reduce the training time needed to learn to interact with complex systems, plus allowing a faster response time on the part of the human, which may be especially important in critical situations.

When considering the use of speech in the context of system control, many advantages are frequently cited. As well as the perceived naturalness of speech, other benefits include the usefulness of speech in the 'hands/eyes busy' and information overload situations, when other channels are stretched to capacity or exhausted, as well as the increased mobility that speech input and output affords the user who is free to move around the workplace. With the use of a radio link, it is possible for information to be collected and received without having to be in close proximity to a host computer. This mobility can be further extended by the use of portable speech recognizers and synthesizers, which workers can use in the field for remote data-entry (Dukarm, 1987). Speech recognition allows data-entry to be direct to a computer system with minimal interruption to the user's primary task. As a result, it is often carried out with greater accuracy, because a step in the data processing cycle has been removed, i.e. collection and input of data occur simultaneously. The attention-grabbing property of speech messages may prove important in some applications, e.g. on the flight deck to attract the crew's attention immediately to a situation that is becoming critical. However, this feature may also prove to be disadvantageous, since 'talking' objects, e.g. lifts, cars, video recorders, have been found to be irritating and a source of annoyance to users (Potosnak and van Nes, 1984). Finally, there are specific benefits relating to specialist groups such as disabled users. Engelhardt *et al.* (1984) cited the following advantages of speech recognition in this context.

1. Speech is sometimes the only remaining means of communication the individual has.
2. It requires no physical linkage between the user and the recognizer, although in some situations, a microphone lead may be needed – this results in remote (and therefore hygienic) control.
3. It allows the disabled user the maximum degrees of control freedom.
4. It can be personalized to accommodate the individual needs of the user.

All these benefits could be applied to speech output, which has been successfully used with vocally impaired individuals to provide them with a 'voice'.

Although the naturalness of speech is accepted in human interactions, it is debatable how 'natural' speech is for human–machine communication. In addition, the transitory nature of speech means that it must be attended to immediately, otherwise retention of the content of the speech output is quickly lost. It is known that the comprehension of spoken messages takes longer than when reading written text of a similar complexity, with the amount of time needed increasing as a function of the speech quality (Baber, 1991a). Although it could safely be concluded that there are many benefits to using speech input and output, there are also specific problems relating to the technology, which impair and limit the performance of speech recognition and synthesis systems, and these are discussed in more detail later in the chapter.

Speech recognition

Brief history

The concentrated research effort resulting from the demands of the Second World War triggered off many electronic developments which indirectly enhanced and accelerated work in the field of speech recognition. Work on Automatic Speech Recognition (ASR) began in earnest in the post-war period when claims to have developed the first speech recognizer were made. In 1952, a device was demonstrated which had the capability of recognizing the 10 digits when they were spoken singly into the machine (Lea, 1980). Recognition accuracy was good, and the machine correctly recognized 97–98 per cent of the digits. One drawback was that in order to obtain this level of performance, the equipment had to be specifically set up for each speaker. This type of system today would be classified as speaker-dependent. This device was experimental, and it is thought the first operational use of automatic speech recognition was in 1973, when the Owens-Illinois Corporation introduced voice input into the inspection tasks of faceplate components during television manufacture (Martin, 1976).

Four decades later, the goal of communicating with machines using human speech has still only been achieved with limited success in terms of recognition performance, and usability. The large part of the research drive has been towards developing more sophisticated algorithms for speech analysis and classification in order to improve machine recognition of speech, e.g. the development of 'dynamic time warping' (where mathematics are applied in order to stretch and squeeze the pattern of the incoming utterance to obtain the best fit with each of the templates available) and 'hidden Markov modelling' – a technique for capturing (statistically) the variations in the way a speaker pronounces a word. Improvements focus on increasing the speed and accuracy of the recognition response, and the size of the vocabulary, while reducing the enrolment time, i.e. the time taken to set up or 'train' the system for a specific user. The latter is particularly important in natural language applications where vocabularies are likely to be in excess of 10 000 words (Billi *et al.*, 1986).

The 1980s witnessed a growing interest in the use of speech recognition, and this is largely explained by the success of the technological advances in the pattern-matching algorithms for recognition, and the increase in UK and European Government-funded research activities. More sophisticated algorithms, plus enhanced computer processing capabilities accompanied by falling costs and greater awareness of the possibilities attainable with human–machine speech communications, all help explain the expanding interest in speech recognition.

Current technology

Recognition of human speech

Present-day speech recognizers use algorithms or mathematical models designed to handle human speech utterances, which may be non-speech sounds (grunts, coughs, uhms, etc. often referred to as 'babble'), phonemes, words, or phrases. A generalized form of the sequence of events is presented below - speech recognizers work on similar principles and this is an adaptation from Simpson *et al.* (1985).

Conversion of the human utterance

The first stage is to convert the speech utterance from its analogue form to a digital representation. Various techniques are used in order to achieve this, e.g. filter banks may be used or various time series analyses or linear predictive coding. The end-result is a compression of the data to reduce the memory space needed.

Normalization of this data

These digital data are normalized in order to provide a control for the various speech rates, amplitudes, pitch levels, etc.

Feature extraction

Any acoustic features that are known to represent different utterances will be extracted here.

Pattern-matching

The digitized form of the incoming utterance is compared with templates that have been previously stored. Template storage may have been during the initial system set-up or during updating of the templates while the system was in operation.

Decision-making

An attempt is made to locate the best match between the utterance and the stored templates. Depending on the pre-programmed selection criteria, this will result in either a 'recognition' if the fit is good, or if the system cannot find a suitable match, a rejection, i.e. an unrecognizable utterance. The various limits for this final part of the recognition process will have been previously decided by the system's designer. It is this stage which takes up the processing time, which increases with the size of the vocabulary.

Omitted from the above recognition process is any reference to the problem of end-point detection of the speech utterances. In everyday speech people do not talk with pauses between the words, and unless the application warrants the input of isolated words, the recognizer has to be designed to cope with co-articulation.

Although principles of speech recognition have been developed, speech recognizer devices vary according to a number of parameters. These centre around the type of speaker, the vocabulary used and the demands of the task, and include the following.

Speaker-dependence vs. speaker-independence

Speech recognition devices are generally categorized into two types: speaker-dependent and speaker-independent. The difference concerns whether the speech utterances of the user are known to the system. Speaker-dependent devices require the intended user to provide samples of the words to be used - a process known as enrolment, while speaker-independent systems in theory can recognize the utterances of any user, whether or not known to the system. Enrolment is achieved by repeating the words in the vocabulary a number of times, although some manufacturers of commercially available devices state that their product only requires single pass enrolment, i.e. one utterance of each word. However, it is generally thought that a single utterance will not capture sufficient of the variation in speech to provide adequate information (Baber, 1991a). Baber concluded that in order to trade off information content and time, 3–5 utterances would be sufficient for each word. Set-up times could therefore be quite long, as a vocabulary of only 50 words would require up to 250 spoken samples to be provided. The alternative approach is to use speaker-independent technology, which does not require enrolment as the vocabularies are pre-stored.

In Baber's review of the commercially available speech recognition devices, he concluded that the current state of the art is dominated by speaker-dependent devices (Baber, 1991a). Extrapolating from his figures, around 60 per cent of commercially available devices are speaker-dependent. However, it should be noted that this state of affairs is changing with an ever-increasing percentage of the market being taken up with speaker-dependent recognition systems. To quote Baber, 'in 1983, speaker-independent devices accounted for 77 per cent of the market while in 1989, they accounted for 60 per cent'. Recent trends have seen a shift towards speaker-dependence, and looking to the immediate future, it could be predicted that this will continue, due to the scale of the technological problems which have to be overcome in order to achieve acceptable recognition performance with speaker-independent systems.

Vocabulary constraints

Besides providing speech samples, the major difference between speaker-dependent and independent recognizers is the extent to which the vocabulary can be designed and tailored to suit a specific task domain. As already stated, speaker-independent devices come with pre-recorded speech samples, which means that the decision as to what constitutes an acceptable vocabulary has been made before purchase of the device. In general, these recognizers employ a small vocabulary of less than 40 words usually comprising commonly-used command words and the digits 0–9. Speaker-independent devices may be useful in enquiry-based systems, and quality control/inspection tasks, which have either a small vocabulary and/or a rigid syntax accompanied by a limited vocabulary requirement. They tend to be inappropriate if required for tasks with either large or specialized vocabularies.

Performance as measured by recognition rates relates to the size of the active vocabulary. (This is well known to sales people who when demonstrating speech recognition technology achieve spectacular recognition performance largely due to having a limited vocabulary). In our own work with a vocabulary of the digits 0–9 plus some command words, we have achieved recognition rates across a population of users of over 95 per cent (Frankish and Noyes, 1990). One characteristic of successful, operational applications of speech recognition is the employment of only small vocabularies, as current technology is inappropriate for coping with the demands of the vocabularies needed for natural language processing. One way to cope with large vocabularies is to introduce syntax, so that only a small part of the vocabulary is active at any one time. For example, Hill (1980) reported that 13 000 distinct sequences could be performed with a vocabulary of just 54 words in a fighter aircraft application. The benefit of this approach is that the recognizer can handle a large vocabulary without the accompanying degradation in recognition performance. With respect to vocabulary design, there are other ways of improving recognition performance, e.g. the use of polysyllabic words or short phrases has been shown to enhance recognition rates, while avoiding the use of acoustically similar words (Noyes and Frankish, in press). It is possible, therefore, for prudent vocabulary selection and design to aid optimization of recognition performance.

Isolated and connected word recognition

Speech recognition devices can handle speech in one of two ways: isolated word recognizers require each word to be spoken in isolation, i.e. with a pause between each word, while connected word recognizers can cope to a limited degree with words spoken in a continuous stream. Here the pauses between words are shorter (but not eliminated), and a talking speed which users find acceptable is possible. Connected word recognition is often taken to be analogous with continuous word recognition.

Technologically, problems are associated with recognizing the beginnings and endings of words in a stream of speech. The consequence of this is that isolated word recognition tends to be more accurate than connected. There are also other penalties relating to the use of the latter, and which will all have detrimental effects on recognition performance. These were outlined by Noyes *et al.* (1992);

1. humans tend to find it difficult to pause between words, especially when working in stressful and/or high workload situations;
2. individuals find it irritating to have to speak very slowly; and
3. users may alter their normal style of speech, unnaturally accentuating words, etc., when asked to insert pauses.

To conclude, current speech recognition technology is still best suited to tasks which use speaker-dependent, isolated word recognition and small, carefully selected, vocabularies of between 50 and 1000 words. However, world-wide there are many research groups working on solving the converse problem of achieving acceptable recognition with speaker-independent systems handling large, non-specific vocabularies of connected speech in excess of 1000 words. It is likely given this research effort, and the continuing developments in computer processing power, which give systems a greater ability to handle more data with faster response times, that breakthroughs will be achieved in the next decade in attaining speaker-independence. Furthermore, advances in AI (artificial intelligence) techniques modelling natural language and their subsequent incorporation into speech recogniton software will introduce 'intelligence' into speech systems, thus bringing the goal of achieving recognition of unconstrained human speech a little closer. One consequence of developing the natural language aspects in such communications is the increase in expectations on the part of the user. Paradoxically, they may then make unrealistic assumptions about the comprehension capabilities of the system. This could have adverse effects on recognition rates, as the person regresses to colloquial-type language.

Applications

Despite the amount of research work and the wealth of knowledge that now surrounds this topic, there are surprisingly few actual end-users of speech recognition systems (Noyes and Frankish, 1987), although many applications are still at the experimental stage. From our own survey work, we located only one fully established and operational application in the UK. This involved the use of speech recognition in the making of maps and charts by cartographers and hydrographers at two different establishments in the South of England. Here, voice input had been successfully integrated into an existing computer-aided design (CAD) system and in operation for over a decade. Employing the speech channel had originally been considered because the eyes and hands of the individual at the drawing board were fully occupied (the 'eyes busy/hands busy' scenario) and utilizing speech constituted one of the few remaining options for entering information (soundings, place names, land features, etc.). Speech recognition has been introduced at various stages in the map making process, but its primary use is in co-ordinating the position of the symbolic and textual information as it will appear on the final version of the map. In order to achieve this, the user aligns a cursor with cross-hairs onto the appropriate location, which can be detected by a matrix of wires in the table, and this information together with the spoken feature is relayed to the host computer. The obvious alternative in this application would have been to enter data via a keyboard, but this would have resulted in the user continually having to interrupt the task in order to key in information.

In the USA, the situation is very different as industrial speech recognizers have been in widespread use since the mid-1970s (Martin, 1976), primarily in quality control and inspection tasks. In manufacturing industry, it is common practice to inspect finished products for defects or abnormalities, and often an inspector has to enter information about products into a computer system at the same time as they are being examined. The installation of a speech recognizer provides a medium whereby product examination (and perhaps various measurements) can occur simultaneously with verbal data-entry. In this and other similar quality control tasks, introducing speech recognition resulted in a time saving plus an increase in accuracy (Martin and Collins, 1975). Such benefits accrue because reducing the number of steps in the data handling process results in there being less opportunity for the information to become corrupted, especially if defects were being noted manually before input to a computer system.

Materials' handling is another application area where speech recognition has been successfully implemented with one of the first uses being in Europe in 1974 (Rehsöft, 1984). A voice recogniton unit was introduced into a system controlling the movement of freight along conveyor belts. The operators checked the address of each sealed carton for its destination, and then spoke the appropriate conveyor spur number followed by a command word into a microphone. Speech recognizers are well suited to this and similar tasks, because a small vocabulary of under 100 words is needed and misrecognitions do not result in critical situations, merely a package having to be retrieved or re-routed. Errors are inevitable with any sorting system, and it is likely that the faster sorting rates achieved with speech recognition will compensate for any error increase, if it occurs. However, in some sorting applications, speech recognition technology has indicated a need for fewer staff to complete the job. Specific examples where this situation has arisen include American airport baggage handlers directing luggage (Lind, 1986) and postal workers sorting parcels (Martin, 1976). Sorting tasks usually involve two employees: a 'facer' to turn the parcel around, and a second person to key in the information concerning the package's destination. Introducing a speech recognizer reduces this workload and results in only one person being needed to complete the task. One of the ramifications from this reduction in personnel is that speech recognition is not always viewed as being a welcome addition to the workplace.

In contrast to the above operational applications, there are a number of situations where the implementation of speech recognition is being contemplated and pilot trials have been conducted. Again, the common feature is a person attempting to complete a task while having to process several pieces of information simultaneously. One example is the use of speech recognition in medical applications (Confer and Bainbridge, 1984). In the USA, ophthalmic surgeons have found it advantageous to control pieces of equipment using voice input during eye operations. Sterility requirements proclude using the hands to control surgical microscopes, and foot controls have been found to be difficult to operate and error-prone, especially in a darkened theatre. Speech recognition technology allows the surgeon to focus more fully on the primary task of carrying out the operation without having to stop to instruct a non-scrubbed assistant to adjust the equipment.

Other application areas where speech recognition is under serious consideration include avionics, the office environment, telecommunications and for disabled users. Each of these make very different demands on current technology, and because it is difficult to draw definitive conclusions across this group as a whole, they are considered separately.

In aerospace applications, the ever-increasing complexity of airborne systems continues to increase pilot workload. This has resulted in other communication channels having to be considered including the viability of speech input. The benefits of utilizing speech recognition in this context have been covered in detail by Starr in Chapter 9, although to date, the precise nature of the involvement of the recognizer in airborne management systems has not been well defined. This is partly because the research effort has focused more on overcoming the more immediate problems of obtaining acceptable recognition performance in this particular physical environment with high noise, and vibration levels (and high G forces in the military cockpit), and possible changes in voice patterns due to psychological factors such as extreme stress and high workload.

Automation of office activities is an area which attracts a great deal of research interest, and it has been estimated that office workers spend between 15 and 20 per cent of their time in verbal communication. If some of this already existing speech could be captured and acted upon automatically, it is hypothesized that financial savings and increased worker productivity might be achieved. For example, in data-entry tasks, speech input is likely to be faster than keying, especially for the non-skilled and semi-skilled keyboard operator. The main thrust of the research in this area has considered the use of speech input for text creation in word processing and voice messaging. However, both these applications along with the majority of data-entry and information retrieval tasks place heavy demands on the capabilities of the speech recognizer. In order to make these office applications workable, a large vocabulary in excess of 5000 words will be needed (Mulla, 1984); as already stated, present speech technology lacks the processing power, although it could be predicted that this obstacle will be overcome in the future. A further issue concerns the use of isolated versus connected word recognition. The former type of system currently achieves better recognition rates, and so would be anticipated to be more appropriate for operational applications. However, it is unrealistic to expect office workers to input large chunks of text with slight pauses between each word, and therefore, isolated word recognition is not feasible for these large vocabulary applications. One solution to overcome the limitations inherent in the recognizer technology has been the suggestion of a multi-modal system for office applications. For example, all editing and formatting commands could be given verbally, while text is entered via the keyboard. This approach has been adopted by some researchers (Morrison and Green, 1982; Leggett and Williams, 1984), but the benefits to the user are not immediately obvious. When considering the indices of speed, accuracy, ease of use, there appears to be little advantage in introducing speech to the system. But, this does not take into account user preference, which may favour this dichotomy of operation. Future work may indicate individuals like 'verbal editing', because, for example, it is less disruptive to the primary task of text input.

The use of a speech recognizer in conjunction with telephone activities is seen by many as an attractive proposition, and this area has been under investigation for many years (Gibson and Bruce, 1984; Willis *et al.*, 1984). One specific example is the use of speech recognizers to replace the human operators in telephone information and enquiry services (Tagg, 1988). Total automation of these systems is unlikely, since there will inevitably be situations when control will have to revert to the operator. These may be due to technical problems or simply a preference as stated by the caller to talk to a fellow human. Implicit in the telecommunication applications is the need for speaker-independent systems, since user enrolment is not a practical proposition. In addition, there are other considerations affecting recognition performance and which arise from dealing with a large vocabulary from a group of unknown speakers eliciting a variety of dialects and accents, plus the added problems of the recognition system handling intonation, prosody, gender differences, and an individual's variations in speech over the course of a dialogue.

There is one very diverse group of individuals for whom speech recognition may have a major role to play in enhancing the quality of their lives, as for some disabled people, speech input provides one of the few remaining options allowing them to communicate with fellow humans, and have some degree of control over their environment and transport needs. As a group they tend to benefit from research developments in other areas, and results from specific studies with disabled users have been encouraging (Noyes *et al.*, 1989). Trials have been carried out with voice-activated environmental control systems (Haigh and Clarke, 1988), wheelchairs (Youdin *et al.*, 1980) and robotic arms for carrying out everyday domestic-related activities (Curtis, 1988). The advantage of speech recognition in this context is that users have only to maintain consistency in their speech, as this is the basic premise upon which current recognizers work, and need not necessarily be fluent speakers as defined in terms of human to human communication. Again, one of the limiting aspects of the technology concerns being able to achieve acceptable recognition performance. However, this tends to be counter-balanced by the high motivation levels of this user group, who have been shown to tolerate poor recognizer performance as it may provide their only means of completing a task independently (Haigh and Clarke, 1988).

To sum up, the characteristics of successful applications include having small vocabularies of less than 100 words, the ability to use isolated word recognition, a dedicated user group, and a lack of other options providing as desirable a means of completing the task. With regard to the latter, speech recognition may be viewed as a last resort in this context, and this may apply in some industrial and disabled user applications. Technological advancements in the future will inevitably 'convert' some of the currently experimental applications to operational ones, and speech recognition technology will become a common-place feature of the office environment and in telecommunications.

Speech synthesis

Brief history

Historically, attempts to reproduce human speech preceded the development of devices to recognize human speech. The imitation and reproduction of human speech has been a source of fascination for many centuries and early attempts at duplicating human speech included Kratzentstein's vowel synthesizing machine and Kempelen's talking machine in the late eighteenth century (Sclater, 1982). Another milestone was the development of VODER (Voice Operation Demonstrator) at Bell Laboratories, USA in the 1930s, which was the first electric voice synthesizer to produce connected speech (Bristow, 1986). A further breakthrough occurred in 1978 with the introduction of the first single-chip synthesizer, which was incorporated into a child's educational toy 'Speak and Spell' (Schalk *et al.*, 1982).

Unlike speech recognition, there are many different ways of generating human speech and several categorizations of the various processes have been devised. One such classification was described by Hollingum and Cassford (1988), who divided speech generation into the following three types:

1. the reproduction of fixed messages, as in the tape-recording and playing back of a message;
2. the production of variable messages – the earliest example of this is TIM, the speaking clock, introduced in the 1930s; and
3. the development of synthesized speech.

The first two types of speech output can be obtained by using pre-recorded human speech; this introduces an element of inflexibility into some systems, as all messages must first be recorded using a human speaker. On the other hand, they may be wholly appropriate for applications which employ a restricted set of predictable words or phrases. In terms of current technology and applications, it is the third area of speech synthesis which is of more interest, and will be described in detail here.

There are two general approaches to the generation of synthesized speech: speech synthesis-by-analysis and speech synthesis-by-rule. Synthesis-by-analysis uses human speech as the basis for producing digitized computer speech with the pitch, voice quality, and formant characteristics of the speech being analysed in order to produce a speech output, which sounds very similar to human speech. Like the first two categories, described by Hollingum and Cassford, this approach has limitations in that the vocabulary has to be encoded before use, and the storing of only a relatively small vocabulary takes up a proportionally high amount of computer memory. Baber (1991b) concluded that although digitized speech may be useful in simple applications such as toys, it is unlikely to be appropriate for more serious use in industrial applications. In contrast, synthesis-by-rule speech is generated totally by rules derived from linguistic and acoustic parameters and without using an original recording of human speech. As a result, this type of synthesized speech has greater flexibility in that it allows messages to be constructed in response to local demands and different situations. Current research activity in this area is towards improving the quality of the synthesized speech output. State-of-the-art technology is capable of producing highly intelligible speech, but it does not always sound natural (Bailey, 1985). As a general rule, speech output using synthesis-by-analysis techniques tends to attain a much better voice quality than the synthesis-by-rule systems. This is partly due to the inherent difference in the two processes with the latter attempting a more complex approach in the production of synthesized speech. Difficulties arise because of (*a*) the absence of well formulated prosodic rules (without which 'meaning' cannot be transmitted to any great extent) and (*b*) the absence of acoustic variability modelling, which is intrinsic to human speech (Sorin, 1991).

Current technology

Generation of human speech

The relative simplicity of the technology needed for the first two types of speech output as defined by Hollingum and Cassford (1988) does not warrant further discussion here. The research effort and emphasis is currently on text-to-speech synthesizers with speech being generated using synthesis-by-rule techniques; this area will provide the focus for discussion.

A 'simplified' series of events in text-to-speech synthesis would be as follows: this is largely adapted from Waterworth and Holmes (1986) and others.

Text-to-synthesizer processing

The message to be synthesized is derived from printed text by an OCR (Optical Character Reader), or directly from an ASCII (American Standard Code for Information Interchange) coded text file.

Normalization of text

The text is normalized to convert all tokens into full alphabetic forms, e.g. 'Mr' into 'mister', '1st' into 'first', etc.

Phonemic analyses

A corresponding sequence of phonemes or diphones (pairs of phonemes) is computed together with supra-segmental (prosodic) information about stress placement, pitch and speech rate. Words and parts of words are matched to a dictionary of morphs held in the system. Allen (1980) suggested that as many as 95 per cent of words from randomly selected text can be successfully identified using a dictionary of just 12 000 morphs.

Use of exceptions dictionary

The exceptions dictionary is essentially a look-up table used to identify the pronunciation of known words and parts of words. If a word from the text cannot be identified from the look-up table, letter-to-phoneme rules are used to try and obtain the pronunciation. Current commercial text-to-speech devices hold about 5000 rules which are used in combination with around 3000 items in the exceptions dictionary.

Output of speech

The use of interpolation rules optimized for speech quality are applied and the appropriate sequence for the sound pattern of the message as a whole is output. Although the synthesized speech output from this type of system is usually highly intelligible, it does not always sound natural – the so-called 'computer accent' of most synthetic speech. As a result listeners have to put more effort into processing synthetic speech compared to natural speech. This aspect of synthesized speech is well documented in the human factors literature (Luce *et al.*, 1983; Pierce and Remington, 1984; Waterworth and Thomas, 1985).

Text-to-speech conversion is largely a mechanical activity carried out independent of meaning and to a large extent grammar (Waterworth and Holmes, 1986), and as a result, there are shortcomings in the quality of the finished product. The three parameters often used to evaluate speech synthesis systems are data rate, intelligibility and naturalness.

Simpson (1983) stated that the meaning of the term **data rate** when referred to in the speech synthesis literature could be confusing, because of different interpretations. It can refer either to the amount of storage needed for the speech data, or the rate of transmission of the speech data, or the rate at which the actual synthesized speech is presented. The synthesis-by-rule approach tends to be efficient in terms of memory requirements but because it is computationally demanding, the whole process is slow. From a human factors point of view, the rate of presentation of the synthesized speech could be critical, as users are likely to become frustrated with very slow rates. In contrast, human listeners are able to maintain understanding of compressed speech, i.e. normal speech delivered at a fast rate.

A definition of **intelligibility** in this context would be the percentage of speech units correctly recognized by a human listener, where the units could be phonemes, words, sentences, etc. All synthesizers have to obtain a baseline level in terms of intelligibility in order to be functional, and experimental work with users has shown that it is possible to define the rules relating to the various processing stages for making speech intelligible (see, Nusbaum and Pisoni, 1984). The degree of intelligibility required in a particular system is greatly dependent on the application, as it is imperative in some situations that intelligibility is high in order to achieve understanding of the entire message, e.g. giving orders in an emergency. A general conclusion is that it is possible with current technology to achieve intelligibility levels similar to those obtained with human speech.

In contrast to intelligibility, **naturalness** is a subjective measure, and refers to a listener's judgement of the degree to which the speech sounds like human speech. There are no standardized tests of naturalness (Simpson, 1983). The degree of importance of how natural the speech sounds will depend upon the task and the user, and it is difficult to make any generalizations.

Together intelligibility and naturalness will determine the quality of the speech output. Defining speech quality is not easy as it implies some measure of acceptability on the part of the listener (Lienard, 1980). However, the quality itself may not actually be important as it may be sufficient that the synthesized speech provides a verbal message immediately obvious as being different from a human speaker, e.g. the use of a female voice in the aircraft cockpit for alerting purposes. Furthermore, a poor quality speech synthesizer may not necessarily result in a poor understanding of the spoken message, as intelligibility and naturalness are not necesssarily correlated. It is possible to have natural sounding speech which due to background noise may be totally unintelligible. Conversely, some highly intelligible messages may be conveyed via a robotic-sounding, monotone voice. In terms of speech synthesis, both these systems may be labelled by users as generating poor quality speech.

To conclude, unlike speech recognition there are a greater variety of ways of generating speech output. This increased range of options provides the designers of potential applications with (a) a selection of speech output techniques from which the most appropriate can be sought, and (b) fewer limitations on the use of speech synthesis, e.g. there are no constraints in terms of vocabulary size on applications since synthesizers can handle a few recorded messages through to unrestricted output. However, in a text-to-speech system the penalty for using the latter type of output is likely to be a degradation in speech quality. Like most computer systems, trade-offs in the features selected exist, and given a fixed memory capacity, the quality of the speech will be determined by vocabulary characteristics (Schalk *et al.*, 1982).

Applications

The variety of ways of achieving speech output plus the relative simplicity of the problem being addressed has resulted in a number of well established applications. A common feature of these applications is that speech synthesis tends to be an 'add-on' feature to a device as opposed to being an integral part of a system. Consequently, speech output often has only intermittent usage (and indeed, there may be occasions with some speech synthesis applications when the audience does not exist at all). In the case of speech synthesis, as opposed to recorded speech, this may be partly explained by the fact that humans find it tedious to listen to large chunks of synthesized speech. In contrast, speech recognition applications often assume a control function which demands operation on a more continuous basis. This may explain why the characteristics of the user population are more important in the design of speech recognition systems.

Speech output applications may be broadly categorized according to use. Speech synthesis has been employed to:

- offer guidance, and give warnings;
- issue instructions;
- 'speak' written text;
- provide feedback.

It should be noted that there is a degree of overlap between some of these application areas, and it is impossible to be definitive about speech synthesis uses.

The use of speech merely to provide guidance is not commonly found as usually systems of this type will emit warnings, rather than just guidance. One example of a system only providing guidance information would be in the case of lifts announcing floor numbers. Synthesized warning messages are usually associated with transportation applications. Some examples include: vehicles that issue safety warnings, e.g. seat belt reminders; undergound trains that remind passengers 'to mind the doors'; the use of warning mesages on the flight deck, e.g. 'retard, retard' on the A320 aircraft, to remind the crew to pull up. There is not a clear dividing line between warning an individual and giving them instructions, and the latter example in some situations will act as an instruction rather than a warning. A further example might be automated public-address systems providing information about evacuation procedures in the event of a fire. Often this use of speech output has received adverse publicity, primarily because of the 'irritation factor'. It could be hypothesized that the degree of annoyance is proportional to the length of the synthesized speech with individuals being more tolerant of short messages. Many of the consumer products which feature speech synthesis warnings tend towards the gimmicky utilizing the novelty aspect, e.g. washing machines that have been programmed to announce that the washing cycle has finished. Despite this aspect, there are likely to be users from the disabled population who will benefit from 'talking' domestic appliances.

The second category of speech synthesis use has been to provide verbal instructions. Applications include: setting up video recorders and equipment in industry, and display screen-based activities, e.g. spreadsheets with verbal prompts for setting up columns, etc.; carrying out banking transactions at terminal points; and educational uses. The latter is an area which has attracted research interest over the last decade in line with the advent of the PC (personal computer) in the late 1970s, and the move towards cheap computer technology and its subsequent introduction into schools, and colleges. In educational applications, there is a large degree of overlap between the second and third category in that systems which use synthesized speech instructions will also provide speech output of written text. This is a well-researched area and systems like the 'talking typewriter', the 'language master' and the 'talking text writer' have combined word processing with synthesized speech in order to provide the beginner reader with the opportunity to turn written words into speech (Rosegrant, 1986). An extension of this concept is the use of speech synthesis in CAL (computer-aided learning) systems designed to help the beginner reader, e.g 'talking books' (Davidson *et al.*, 1991).

An underlying theme across all the speech synthesis applications is their relevance to some groups of the disabled population. Although there are specific prosthetic devices for individuals with speech disabilities, the use of synthesized speech in the context of warnings, instructions and feedback is of immense benefit to visually handicapped people. One recent example of the third category has been the development of 'talking newspapers' for the visually handicapped (*Guardian*, 11th March 1993).

The fourth category is the use of speech output to provide feedback. In some applications, e.g. remote data-entry, and some educational applications, speech feedback will naturally follow speech instructions given by the computer, or speech recognition commands to the system. This approach is supported by Wickens *et al.* (1984) in their theory of multiple resource, which essentially stated that stimulus-response compatibility should be maintained in system design. The implication from this is that speech output should follow speech input.

A comparison of speech synthesis and recognition application areas indicates that the nature of the use in terms of criticality is not as important in speech output. This is due to the more passive role assumed by speech synthesis as an information provider, whereas speech recognizers are totally reliant on the recognition process being successfully carried out. As a result, speech output is appropriate and indeed operational in avionics and disabled user applications in contrast to speech input, where misrecognitions by the recognition system cannot be tolerated.

The future

The development of speech recognition and synthesis techniques are still in their formative years, and to date no single approach has emerged as universally acceptable for either. However, with judicious and careful selection of applications, both types of speech systems can be used successfully with current technologies. When considering speech recognition, further developments will have to overcome the problem of misrecognitions, whereas with speech synthesis, the research effort will be towards improving the quality of the speech output. The implications of each of these are considered in turn.

Some individuals may consider it ironic that we ask for flawless recognition between humans and machines, when human to human communication is far from perfect. However, there are some application areas where the well being of the users may be threatened if their verbal input is not correctly recognized, e.g. in the airborne environment and for some disabled user applications such as wheelchair operation. In this context, the problem of misrecognitions is hampering further take-up and developments in these areas. Although the attainment of 100 per cent perfect recognition is an unrealistic goal, it could be predicted that over the next decade advances in the technology, and more awareness and knowledge concerning the benefits that arise from human factors engineering of systems will contribute to improving recognition performance. With respect to the former, technical advances predicted by Furui (1991) included;

- the development of mathematical models to imitate speech production and perception mechanisms;
- greater understanding of the linguistic processing methods needed to generate free conversational speech; and
- adaptable recognition algorithms to cope with speaker variations (e.g. new speakers) and changes in the environment (e.g. background noise).

He also mentioned the importance of further work on the human–machine interface. In human factors terms, the following can all contribute towards increasing recognition rates to more acceptable levels, and require further research:

- well planned enrolment procedures to facilitate ease of training the recognizer;
- dialogue design (e.g. careful selection of commands, the use of confirmation phrases, the introduction of syntax and 'intelligent' software);
- appropriate error detection and correction protocols;
- feedback mechanisms; and
- techniques for reversion.

On a more global note, future recognition systems will have moved from small sets of known talkers using limited vocabularies towards the development of recognizers that attempt to understand connected speech from groups of unknown speakers. This will inevitably lead to an increase in the number of applications for speech recognition, and based on previous trends, production costs will continue to fall, and the advent of the single chip recognizer will result in the novelty angle of speech devices being exploited. This will lead to an increase in home and toy associated applications with perhaps a trend towards personalized, and customized systems. This hypothesized ubiquity combined with the aerospace research effort concentrating on minimizing recognition errors will indirectly benefit the disabled user group. As a result, they will have increased access to cheaper speech recognizers with improved performance levels. Although the use of speech recognizers will continue to multiply in the short term, it will be many years until the technology is sufficiently reliable to cope with some of the office and telephone applications (e.g. teleconferencing, teleshopping, teleworking) which require large vocabularies, continuous recognition and natural language processing. In the long term, two specific areas which are likely to benefit from these technological advances are speaker identification and verification applications. To crystal ball gaze, it could be predicted that such developments will not take place until the 21st century.

One area where developments are likely is in speech understanding systems. Information concerning different knowledge sources, e.g. the speaker's speech style, topic of conversation and the circumstances in which the dialogue is taking place, will be utilized to co-operate in the recognition process and this will be a key feature in the development of more sophisticated models of human speech recognition, as predicted by Bailey (1985). Advances in speech understanding techniques will decrease the problems associated with the use of large vocabularies, and lead to major improvements in the quality of synthesized speech, and in particular, improving the restricted intonational repertoire of current synthesizers, and the options possible, e.g. the ability to generate regional accents.

Greater use of speech synthesis will not be as apparent as speech recognition, because of the current, more frequent occurrence of operational output applications. It is likely however that the novelty angle will continue to be exploited with increased use in toys and domestic goods. In addition, other application areas, e.g. watches/ clocks, supermarket checkouts, announcement systems, talking terminals, will continue to incorporate speech output until it becomes a commonplace feature in our everyday lives.

Looking to the future, the concept of the 'total speech system' might prove a viable and desirable proposition. In many applications, it might be sensible to employ speech recognizers in conjunction with speech synthesis as in the case of some office and telecommunication applications. For example, electronic filing and voice messaging where information is stored or transmitted in the form of speech, and telephone enquiry systems, where the whole transaction is carried out using speech input and output. The total speech system may also lead to new applications not currently thought of. However, user acceptance may be a problem here. Our own studies have indicated that subjects dislike having a totally speech-based system, primarily because of the ephemeral nature of speech. Despite the greater compatibility of speech input and output over other options, this characteristic of speech many limit greater uptake of speech-based systems. A further problem is that individuals when subject to synthesized speech feedback have been shown to mimic the speech, which may distort the speech recognition input.

One conclusion might be that speech-based technologies are well established enough, they are here to stay. However, in order to enhance and further the use of speech recognition and synthesis both from an application's and users' viewpoint, further consideration and research is needed, and the following provides a list of areas where future work will be focused:

1. a more selective use of speech input and output technology with a growing awareness of what is feasible and what is appropriate – the development and publication of guidelines may be the solution here;
2. development work on recognition algorithms in order to attain continuous speech recognition;
3. development work on improving the quality of synthesized speech using synthesis-by-rule techniques (Chapter 3 by Tatham);
4. the attainment of increased processing power for speech-based systems, which will improve and increase the range of current systems;
5. human factors engineering, to ensure the quality of users' interactions with these systems;
6. more robust speech systems to cope with environmental changes,and speaker variability, in the case of speech recognition; and
7. the development of standards, including measures for assessing the performance of speech-based technologies.

Over the next few years, research in these areas will lead to increasing the ubiquity of current applications, and a number of new applications. Some key areas where developments will be made might include:

1. telecommunication applications in conjunction with telephone activities;
2. in the office in line with the move towards the 'paperless office' concept;
3. in the airborne environment;
4. the toy and games market; and
5. new uses primarily in the educational context, e.g. language translation, distance learning, multimedia applications.

References

Allen, J., 1980, Speech synthesis from text, in *Spoken Language Generation and Understanding*, J. C. Simon (Ed.), Dordrecht: D. Reidel.

Baber, C., 1991a, Human factors aspects of automatic speech recognition in control room environments, *IEE Colloquium on Systems and Applications of Man–Machine Interaction using Speech I/O.*, London: IEE Digest No. 1991/066.

Baber, C., 1991b, *Speech Technology in Control Room Systems – A Human Factors Perspective*, Chichester: Ellis Horwood.

Bailey, P., 1985, Speech communication: the problem and some solutions, in *Fundamentals of Human–Computer Interaction*, A. Monk (Ed.), **12**, 193–220, London: Academic Press.

Billi, R., Massia, G. and Nesti, F., 1986, Word preselection for large vocabulary speech recognition, *Proceedings of the IEEE International Conference on Acoustics*, Speech and Signal Processing, Tokyo, Japan, 65–8.

Bristow, G. (Ed.), 1986, *Electronic Speech Recognition*, London: Collins.

Confer, R. G. and Bainbridge, R. C., 1984, Voice control in the microsurgical suite, *Proceedings of the Voice I/O Systems Applications (AVIOS) Conference*, Arlington, VA.

Curtis, G. (Ed.), 1988, *Rehabilitation Research and Development Center Progress Report*, Palo Alto, CA: Veterans Administration Medical Center.

Davidson, J., Coles, D., Noyes, P. and Terrell, C. D., 1991, Using computer-delivered natural speech to assist in the teaching of reading, *British Journal of Educational Technology*, **22**(2), 110–18.

Dukarm, J. J., 1987, Voice data-entry in the forest, *Speech Technology*, March/April, 28–30.

Engelhardt, K. G., Awad, R. E. van der Loos, H. F. M., Boonzaier, D. A. and Leifer, L. J., 1984, Interactive evaluation of voice control for a robotic aid: implications for training and applications, *Proceedings of the Voice I/O Systems Applications (AVIOS) Conference*, Arlington, VA.

Frankish, C. R. and Noyes, J. M., 1990, Sources of human error in data-entry tasks using speech input, *Human Factors*, **32**, 697–716.

Fururi, S., 1991, Future directions of speech recognition research, in *Natural Language and Speech*, E. Klein and F. Veltman (Eds), 175–6, Berlin: Springer-Verlag.

Haigh, R. and Clarke, A. K., 1988, Evaluation of a voice recognition system, VADAS, for use by disabled people, Report to DHSS, Royal National Hospital for Rheumatic Diseases, Bath, UK.

Hill, D. R., 1980, Spoken language generation and understanding by machine: a problems and applications oriented overview, in *Spoken Language Generation and Understanding*, J. C. Simon (Ed.), Dordrecht: D. Reidel.

Hollingum, J. and Cassford, G., 1988, *Speech technology at work*. Bedford: IFS Publications UK.

Lea, W. A., 1980, Speech recognition: past, present and future, in *Trends in Speech Recognition*, W. A. Lea (Ed.), **4**, 39–98, Englewood Cliffs, NJ: Prentice-Hall.

Lienard, J-S., 1980, An over-view of speech synthesis, in *Spoken Language Generation and Understanding*, J. C. Simon (Ed.), 397–412, Dordrecht: D. Reidel.

Lind, A. J., 1986, Voice recognition – an alternative to keypad, *Proceedings of Speech Tech '86, Voice I/O Applications Show and Conference*, 66–67. NY: Media Dimensions.

Luce, P. A., Feustel, T. C. and Pisoni, D. B., 1983, Capacity demands in short-term memory for synthetic and natural speech, *Human Factors*, **25**(1), 17–32.

Martin, T. B., 1976, Practical application of voice input to machines, *Proceedings of the IEEE*, **64**(4), 487–501.

Martin, T. B. and Collins, J. C., 1975, Speech recognition applied to problems of quality control, *ASQC Technical Conference Transactions*, 8–15, San Diego, CA.

Mulla, H., 1984, An experimental voice command system for PABX and automated office applications, *Proceedings of Speech Tech '84, Voice I/O Applications Show and Conference*, NY, 2–4 April, 48–52.

Noyes, J. M. and Frankish, C. R., 1987, Voice recognition – Where are the end-users? *Proceedings of European Conference on Speech Technology*, **2**, 349–52. Edinburgh: CEP Consultants.

Noyes, J. M., Haigh, R. and Starr, A. F., 1989, Automatic speech recognition for disabled people, *Applied Ergonomics*, **20**(4), 293–8.

Noyes, J. M., Baber, C. and Frankish, C. R., 1992, Industrial applications of automatic speech recognition, *Journal of the American Voice Input/Output Society*, **9**, 1–18.

Noyes, J. M. and Frankish, C. R., Speech recognition technology for individuals with disabilities, *Augmentative and Alternative Communication*. (in press).

Nusbaum, H. C. and Pisoni, D. B., 1984, Perceptual evaluation of synthetic speech generated by rule, *Proceedings of the Voice I/O Systems Applications (AVIOS) Conference*, Arlington, VA.

Pierce, L. and Remington, R., 1984, Comprehension of synthetic and natural speech in the presence of competing human speech, *Proceedings of the Voice I/O Systems Applications (AVIOS) Conference*, Arlington, VA.

Potosnak, K. M. and van Nes, F. L., 1984, Effects of replacing text with speech output in an electronic mail application. *IPO Annual Progress Report 19*, 123–9.

Rehsöft, C., 1984, Voice recognition at the Ford warehouse in Cologne, *Proceedings of the 1st International Conference on Speech Technology*, Brighton, UK, 103–12.

Rosegrant, T. J., 1986, The importance of speech technology as a learning tool for acquiring beginning reading and writing, *Proceedings of Speech Tech '86, Voice I/O Applications Show and Conference*, NY, USA, 297–8.

Schalk, T. B., Frantz, G. A. and Woodson, L., 1982, Voice synthesis and recognition. *Mini-Micro Systems*, **15**(12), 147–60.

Sclater, N., 1982, *Introduction to Electronic Speech Synthesis*, Indiana, USA: Howard W. Sams and Co. Inc.

Simpson, C. A., 1983, Evaluating computer speech devices for your application, *Proceedings of the 7th West Coast Computer Faire*, West Coast Computer Faire, 345 Swett Road, Woodside, CA 94062, USA.

Simpson, C. A., McCauley, M. E., Roland, E. F., Ruth, J. C. and Williges, B. H., 1985, System design for speech recognition and generation, *Human Factors*, **27**(2), 115–41.

Sorin, C., 1991, Text-to-speech research: technological goals and integration issues, in *Natural Language and Speech*, E. Klein and F. Veltman (Eds), 182–3, Berlin: Springer-Verlag.

Tagg, E., 1988, Automating operator-assisted calls using voice recognition, *Speech Technology*, March/April, 22–5.

Waterworth, J. A. and Thomas, C. M., 1985, Why is synthetic speech harder to remember than natural speech? Paper presented at Conference on Human Factors in Computing Systems, Computer–Human Interaction '85, San Francisco, CA, 6 p.

Waterworth, J. A. and Holmes, W. J., 1986, Understanding machine speech, *Current Psychological Research and Reviews*, **5**, 228–45.

Wickens, C. D., Vidulich, M. and Sandry Garza, D., 1984, Principles of S-C-R compatibility with spatial and verbal tasks: the role of display location and voice interactive display control interfacing, *Human Factors*, **26**, 533–42.

Willis, A. R., Bruce, I. P. C. and Young, S. J., 1984, An experimental database query system using automatic speech recognition over the telephone network, *Proceedings of the 7th International Conference on Computer Communication*, The New World of the Information Society, 30 Oct–2 Nov, Sydney, Australia, 272–7.

Youdin, M., Sell, G. H., Reich, T., Clagnaz, M., Louie, H. and Kolwicz, R., 1980, A voice controlled powered wheelchair and environmental control system for the severely disabled, *Medical Progress through Technology*, **7**, 139–43.

Index

T - #0019 - 071024 - C0 - 234/156/12 [14] - CB - 9780748401277 - Gloss Lamination